ELECTRIC POWER IN VICTORIAN BRITAIN

Nineteenth-Century Science, Technology and Medicine: Sources and Documents

Forthcoming:

Meteorology in Nineteenth-Century Society
4 volume set
Edited by Aitor Anduaga

Nineteenth-Century British Microscopy
4 volume set
Edited by Meegan Kennedy

Nineteenth-Century Natural Theology
4 volume set
Edited by Adam R. Shapiro

The Science of Mind in the Nineteenth Century
4 volume set
Edited by Stephen T. Casper & Jon D. Goss

Science and Sound in Nineteenth-Century Britain
4 volume set
Edited by Edward J. Gillin

Scientific Methodology in Nineteenth Century Britain
4 volume set
Edited by Charles H. Pence

Women in Medicine in the Long Nineteenth Century
4 volume set
Edited by Claire Brock

Insects in Victorian Culture
4 volume set
Edited by John Clark

Electric Power in Victorian Britain
3 volume set
Edited by Nathan Kapoor

Women, Gender and Science in Victorian Culture
4 volume set
Leila McNeill & Anna Reser

ELECTRIC POWER IN VICTORIAN BRITAIN

Edited by
Nathan Kapoor

Volume I
Electric Power Imagined

Routledge
Taylor & Francis Group
LONDON AND NEW YORK

First published 2026
by Routledge
4 Park Square, Milton Park, Abingdon, Oxon OX14 4RN

and by Routledge
605 Third Avenue, New York, NY 10158

Routledge is an imprint of the Taylor & Francis Group, an informa business

British Library Cataloguing-in-Publication Data
A catalogue record for this book is available from the British Library

ISBN: 978-1-032-28164-3 (set)
ISBN: 978-1-032-28167-4 hbk (vol I)
ISBN: 978-1-003-29555-6 ebk (vol I)

DOI: 10.4324/9781003295556

Typeset in Times New Roman
by Apex CoVantage, LLC

CONTENTS

VOLUME I ELECTRIC POWER IMAGINED

Series Introduction viii
General Introduction x
Volume I Introduction xxv

PART 1
Anxiety **1**

1 Reflections on the Decline of Science in England, and on some of its Causes 5
CHARLES BABBAGE

2 What Will He Grow To? (illustration) 9

3 The Coal Question: An Inquiry Concerning the Progress of the Nation, and the Probable Exhaustion of Our Coal-Mines 13
WILLIAM STANLEY JEVONS

PART 2
Imagining Electricity **17**

4 Historical Electricity 21
JOHN BYWATER

5 Progress of the Mind, Canto III 31
ERASMUS DARWIN

6 Electricity 37

7 The Silent Revolution, Or, The Future Effects of Steam and Electricity Upon the Condition of Mankind 41
MICHAEL ANGELO GARVEY

8 Science 49

9 Electricity and the Future 69
F. C. WEBB

10 Electricity in Relationship to Science 77
WILLIAM CROOKES

11 Social Evolution 83
BENJAMIN KIDD

12 The Universality of Electricity 89

13 Prometheus Unbound: Science in Olympus (illustration) 91

14 Telectroscopy 95

PART 3
Experiments **97**

15 On the Electricity Excited by the Mere Contact of Conducting Substances of Different Kinds 101
ALEXANDER VOLTA TO JOSEPH BANKS

16 Mr. Bain's Electro Magnetic Inventions 119
J. C. ROBERTSON

17 On the Production of Insects by Voltaic Electricity 143
ANDREW CROSSE

18 Endless Amusement: A collection of nearly 400 entertaining experiments in various branches of science 149
ANON

19 On Electric Conduction and the Nature of Matter 153
MICHAEL FARADAY

20 Charles Wheatstone letter to Alexander Bain, June 13, 1842 163

21 Industrial Applications 171
HIPPOLYTE FONTAINE

22 On the Augmentation of the Power of a Magnet by the Reaction thereon of Currents Induced by the Magnet Itself 189
CHARLES WHEATSTONE

23 A Dynamical Theory of the Electromagnetic Field 195
MAXWELL JAMES CLERK

24 Aron Meter 205
GISBERT KAPP

PART 4
Self-Reflection-Retrospection "Looking" 209

25 Priority 213
ANON

Index 216

SERIES INTRODUCTION

Nineteenth-Century Science, Technology and Medicine: Sources and Documents

GENERAL EDITORS' INTRODUCTION

This volume is part of the Routledge Historical Resources primary source collection, "Nineteenth-Century Science, Technology and Medicine: Sources and Documents." This series offers a fresh look at the historical development of science, technology, and medicine through the eyes of subject experts and specialists from around the world. The series provides curated access to significant primary-source documents on these subjects, collating and contextualizing these valuable resources so that they may ground not just scholarly research but undergraduate and graduate research and teaching as well. As well as being published in print and ebook formats, the contents of this and all other volumes in this series will be available through the Routledge Historical Resources database, where they will be presented in a fully integrated and searchable format, together with supplemental secondary sources and images and newly-commissioned scholarly written and video essays.

Our editors have framed their volumes around themes they consider most pertinent to understanding nineteenth-century developments in their chosen field and its significance to broader changes in nineteenth-century British culture and society. Often the editors have chosen to provide short extracts to illustrate particular points or themes; in other cases, they have provided entire essays, book chapters, or periodical articles as appropriate. Each volume is prefaced by an introductory essay in which the editors review their topic and discuss the significance of their sources. They also frame each entry or collection of related entries with explanatory headnotes. Thus, each volume or set of volumes will provide a useful resource of easy-to-navigate, well-contextualized sources curated by an expert eye, which will serve as a useful introduction for those unfamiliar with the field. Since the editors build each volume to represent a unique, deeply-considered scholarly perspective on the foundational primary sources for a particular subject, the series should also prove valuable to established scholars, even those with deep experience in the field. Indeed, given the range of topics covered and sources collated in this series, we anticipate that this project will foster an exciting and insightful interdisciplinary perspective on the histories and cultures of nineteenth-century British science, technology, and medicine. We as general editors have learned a lot through our engagement with the editors who have contributed to this project, and we believe that wider audiences will do so, too.

Throughout the project we have asked editors to comment on the state of the existing historiographical debate, and to specify where their own work fits in. Some editors have chosen to do this through the provision of recommendations for further reading; others do so through explicit commentary on this topic in their introductory essays, and in the headnotes and footnotes to each of the sources they have selected.

The histories of science, technology, and medicine are wide and often disparate fields, even though they are often encountered by students in the context of a single university department. Although they have much in common, scholars of science, technology, and medicine also propose field-specific questions and approaches; these are represented in the project through the disparate interests of the range of scholars contributing to the series. We have also worked to bring the insights of literary scholars as well as historians to the project.

We have made every effort to cover major nineteenth-century themes and topics in the scientific and technological disciplines as well as to give space to specialist topics. Of course, as is inevitably the case in a project of this nature, which must rely on the availability of expert scholars in particular fields, there will remain gaps. We have attempted to address these where we can in the form of introductory essays and video essays in the database. We do believe, though, that the reader and researcher will find that this project presents the major themes, as well as many specialist topics, in detail, and with scholarly rigour.

We are very grateful to Routledge for their invitation to take on this project, and we would especially like to thank the commissioning editor Rachel Douglas, her successor, Laura Pilsworth, and Simon Alexander. Most of all, however, we are grateful to the extensive list of scholars from around the world who have agreed to contribute, and who make this series what it is.

Meegan Kennedy and Piers J. Hale

GENERAL INTRODUCTION

During the nineteenth century, the knowledge around and imagination about electricity transformed. Spurred by epistemological shifts in natural philosophy, new conceptions of and applications for energy, and larger communities of people enmeshed within the scientific milieu, electricity became more than an object of curiosity for Victorians.[1] These changes fundamentally altered how people thought about electric phenomena and reconfigured how people conducted, consumed, and ignored science. These changes brought about a reconceptualization of "energy," quantified invisible forces like magnetism, reoriented the possibilities of industrial production, and, quite literally, shed light on new ways of existing in the world. The people who participated in this change included natural philosophers, mathematicians, performers and experimentalists, salespeople, writers, illustrators, theologians, consumers, and marginalized members of the broader Victorian world. Together, they wrestled with electricity, assigned it various meanings, and influenced the foundations of a rapidly electrifying world. By its very nature, electricity branched, writhed, and pulsed through Victorian bodies, technologies, ideas, and infrastructure. By the turn of the twentieth century, it became inseparable from Victorian culture.

Electricity functioned as a vessel by which people imposed new physical systems of regulation, mantras of productivity, and cults of modernity. *Electric Power in Nineteenth-Century Britain* weaves together many of the writings and productions concerning electricity. Given the voluminous number of publications dedicated to electricity in the nineteenth century, the following volumes will explore primary sources concerned with electric power, leaving room for other volumes and experts to cover electric communication technology.[2] In each of these volumes, we will focus on the well-known but obscure works of people concerned with the *utility* or usefulness of electric power. Framing our interpretation of these sources around utility allows us to explore the varied possibilities, realized and unrealized, people imagined for electric power rather than confine ourselves to the eventual electric power systems of the late nineteenth century. Understanding how people envisioned the utility of electricity in nineteenth-century Britain is an essential part of engaging in both British history and the history of science.

A WORD ON SOURCES

Given the parameters of nineteenth-century Victorian Britain, the (in)accessibility of archives, the realities of scientific communication, and my expertise, I chose sources that I believed students and researchers might find the most helpful. The sources in these volumes are not comprehensive nor entirely representative of nineteenth-century work on electric power.[3] They provide a foundation for researchers to conduct more complex investigations into electric power during this period. The sources here showcase not only the well-known periodicals like the *Telegraphic Journal and Electrical Review*, but also less well-known journals, letters, and literary works, as well as well-known publications not specifically concerned with electricity.

As this Routledge series is dedicated to the more extensive history of science in Victorian Britain, it follows that the institutions and presses seen here will also be British. Beyond that more practical decision, Britain functions as an excellent nexus through which to observe shifts in the electrical science worldwide. Declaring one country or culture as the primary focal point of electrical science is impossible and unimportant. The electrical sciences flourished globally during the nineteenth century. They did so both because of the intellectual curiosities and material consumption of massive empires and also because of the extraction of resources, exploitative systems of commerce, and minimalization of peoples that some considered outside the modern world. The British press highlights all of this. Due to the scale of the British Empire, continental politics, and the expansion of American science, many of the major writings on and reactions to electric power flowed through Britain.[4] And I do include work from places outside Britain featured in their press. This also means that the sources here are in English. Again, that is not necessarily representative of many French, Spanish, Japanese, or Chinese works, to name a few, which could be cited in volumes like this.[5]

With a few exceptions, the sources here are textual. Given the constraints of producing such a volume, text seemed the most logical for readers to engage. That said, I included some visuals and imagery to communicate the breadth of electricity's importance to culture. A more complete primary source collection should include the expanse of instruments, lights, motors, cars, and wires that formed the electrical world. And as the volumes are published online, we will have better means of exploring varied media. As researchers use this reader, I hope they can thoroughly analyze and expand the vast array of publishers and authors that documented the diffusion of electrical knowledge and culture. I also hope that this volume can aid scholars in thinking about the limits of the archives I scoured here in future research. What does it mean to transcribe the British electrical engineers' words in the twenty-first century?[6] How should we assign significance to the energy systems they created and the accelerated Anthropocene it caused? How do these archives allow us to see what might have been? Who gets covered up in such records? These questions are not explicitly answered here, but I do believe that a close dissection of the unfolding of electrification must yield answers. If not answers for informing our current energy crisis, then at least a better understanding of our present knowledge and systems so that others might be more well-equipped to ask the right questions.

ELECTRIC "POWER" IN HISTORY

The study of electrical phenomena was not original to the nineteenth century and was not limited to Great Britain. As with so much of the history of natural philosophy, we can likely never identify the earliest observations or ideas about electricity.[7] It features prominently in literature on the origin of numerous cultures; for example, Huangdi, the Yellow Emperor, a semi-divine being in Ancient Chinese literature, could control lightning to do his bidding.[8] Or in the mythology surrounding Zeus in Greece that explained weather, masculinity, and location of the gods on Olympus.[9] In natural philosophical traditions worldwide, people wrote about and interacted with electricity in a number of ways concerning lightning, conduction, and electromagnetism. Aristotle understood lightning, or "thunderbolts," as phenomena borne of clouds as a dry exhalation of air striking the Earth; or Pliny the Elder argued that the lightning that starts fires in the wild is of the same substance that appeared when striking two stones together; throughout the *Canon of Medicine* Ibn Sina used electromagnetic phenomena as a rhetorical device explain the attraction and repulsion of faculties in the body; in *De Erroribus Medicorum*, Roger Bacon used the magnet as a call for "experiment" in making observations.[10]

Certainly, the natural events that we ascribe to electromagnetic mechanisms were categorized by numerous epistemologies outside of the canon of "western" science. Although there may not exist direct relationships to the electromagnetic theory established during the nineteenth century in Britain, we must acknowledge the scope of interpretation of electrical things in the past and present. The Lakota, an indigenous people in the United States, used and continue to use the word *Wakíŋyaŋ* to categorize thunder, but it can also be used to understand and safeguard against celestial beings responsible for thunder and lightning.[11] Māori called lightning uria or as Te Uira, the god or atua of lightning, personified as one of the forces of creation or nature. After the British invasion of Aotearoa-New Zealand, Māori called electrical energy hiko, which could be used to describe both lightning or other electrical phenomena. The example of the Lakota and Māori concepts are but two, there are likely many, many more.[12] They show us that while Western (Victorian) discourse on electricity may be distinct and more transferable to contemporary electrical science, it is not alone. And in fact, it interacted with, influenced, and sometimes, unfortunately, erased knowledge about electricity.

While it is unfair to suggest that serious study of electrical phenomenon did not begin until the seventeenth century, the publication of William Gilbert's *De Magnete, Magneticisque Corporibus, et de Magno Tellure* (1600) provides a useful starting point for these volumes.[13] First, the categories and distinctions he drew between static electricity and magnetic phenomena established the formative debates for the next two centuries. Second, Gilbert's status as an Englishman appealed to many of the electrical thinkers covered in these volumes; many of their texts begin with histories that almost always begin with him. And third, Gilbert's deployment of the word "electric" marks the beginning of the use of the word in the English language. In the text he conducts experiments and argues that the Earth itself is magnetic. He measured electric force and obsessed over the power of the invisible force. Although many of his assertions and theories about electricity came to be disproved, namely that electricity and magnetism are intimately related, his work marks a significant shift in how Western natural philosophers approached electrical and magnetic phenomena. Later in the nineteenth century, many of them would refer to themselves as "electricians." While distinguishing between them and these early philosophers is important, it is useful to denominate the community of philosophers, socialites, entertainers, and tinkerers as "electricians," which I do throughout this volume.[14]

During the seventeenth and eighteenth centuries, the observation and measurement of electrical phenomena fascinated natural philosophers.[15] Early Royal Society meetings featured many dialogues on the interaction between electrical fluids and physical surfaces. On December 7, 1675, Isaac Newton performed an experiment in which he performed a grinding motion with grass to make bits of paper jump. He claimed, "Sometimes also they would leap up in one arch and down in another."[16] The wider scientific interest in standardization and systemization encouraged a drive to identify and explain such phenomena among natural philosophers. In Robert Boyle's *Experiments and Notes about the Mechanical Origine or Production of Electricity* (1676), he utilized glass, feathers, and locks of hair to demonstrate the manipulation of electrical effluvia.[17] Others like Otto von Guericke and Francis Hauksbee developed electrostatic generators or friction machines, usually comprising sulfur or glass spheres that could be rotated against cloth to create static electricity. While not intending to generate electrical energy for consumption or motive power, these devices and subsequent reconfigurations inspired the first electric power generators.[18]

The eighteenth century saw a number of developments in electrical science, the "new science" or, according to Joseph Priestly, "the youngest daughter of the sciences."[19] As important as the development of theory and instruments, electricity became a popular object of discussion outside of the scientific community. It became a thing to discuss, wonder at, and fear. Science came to the salon. Experimenters and socialites such as Benjamin Franklin, Jean Antoine Nollet, Stephen Gray, and Charles du Fay all circulated their ideas for all to see. Lectures and peddlers of the "electric fire" frequented meetings and exhibitions throughout Europe. Light, sparks, and eerie glows filled glass tubes and globes, captivating audiences about the future of electricity. Electricity not only met people's eyes but also flowed through their bodies. Stephen Gray performed the "flying boy" experiment in which someone was suspended via silk rope and then connected to a friction generator so that sparks were created when a guest approached close to his hands or face.[20] Women were featured in several experiments, such as George Matthias Bose's "Electric Kiss," in which a woman stood on an insulated surface after being "charged" by a friction generator. Men from the audience approached after being invited to come to kiss her but found that they could not avoid a painful spark when they got close. Such whimsical performances, while absurd and subject to the worst gender tropes about women's purity or closeness to nature, portend the transformation of electricity into more than just an object of curiosity or entertainment. It became a source of explanation.

Many sought to derive that explanation from quantifiable fact. In step with other schools of Enlightenment natural philosophy, electrical thinkers believed that tools, gauges, and isolated containers would provide more truth about electrical fluid.[21] The number of electrical apparatuses in circulation by the mid-eighteenth century was staggering.[22] To assign more significance to any of them risks reinforcing deterministic ideas about what people expect of electricity. For example, we might mistakenly assume that devices like the Leyden Jar foreshadowed the development of batteries or perhaps even more dubiously that people were thinking of electric power provision. Still, the "Leyden Jar" is illustrative because it is prominently featured in the circles of electrical science. Invented, or at least lengthily described, in 1746 by Peter van Mussechenbroek at Leiden University. Other devices that corroborated the idea that electricity could be captured in a fluid existed in the correspondences of experimenters like Ewald Georg von Leist, Georg Matthias Bose, Daniel Gralath, Abbe Jena Antonine Nollet, Benjamin Franklin, and Johann Heinrich Winckler. More important than the invention of the device is the establishment of a new attitude toward electrical science, best illustrated in a letter by Musschenbroek: "I have found out so much about electricity that I have reached the point where I understand nothing and can explain nothing."[23]

The late eighteenth century saw a brief decline in the seemingly unanswerable questions about the quantification of electrostatics. Electrical fluids remained numerous, and the explanations of what we would call an electro-magnetic field also varied. Instead, by the 1790s, more electricians had garnered interest in animal-medical electricity and electrochemistry. Some physicists and subsequent historians have called this a "standstill" in the history of electrical science.[24] However, the research of this next generation of electricians yielded new questions and answers about electrical phenomena and, more importantly, for these volumes, applications. For example, in 1791, Luigi Galvani published his paper on the excitation of frog legs by electricity.[25] He questioned whether muscle and tissue might possess the ability to accumulate electrical fluid or "animal electricity." Alessandro Volta and Galvani debated whether electricity originated in organic material or needed to be created externally. Galvani devised an electrostatic machine, or accumulator, to demonstrate that what he called electromotive force could be generated using chemical interactions.

In 1800, Volta published the invention of this "pile" of silver, zinc, and cardboard discs. Volta's accumulator signifies the amalgamation of the many ideas of electricity that flourished in the eighteenth century that kicked off the new fields of inquiry during the nineteenth century. The exact quantification of the phenomenon created in a pile remained unanswered. As Volta argued, "this perpetual motion may appear paradoxical, perhaps inexplicable; but it is nonetheless true and real, and can be touched, as it were, with the hands."[26] The pile ended the long string of eighteenth-century instruments and created the opportunity for new ones in the nineteenth century. The study of electricity now belonged to chemistry and physics. The steady current it created helped lay the groundwork for the establishment of the direct relationship between electricity and magnetism.

By the nineteenth century, the study of electricity, and in particular concepts of deriving something from electrical energy, circumvented the world. One could not give a complete picture of these developments from a national or continental perspective. To provide a better context for the sources in this volume, I will specifically focus on the most salient developments in the Victorian sphere. Furthermore, I will explicitly anchor the remaining narrative to electric power. By the nineteenth century, electrical questions obsessed so many different groups of people with an interest in God, communication, chemistry, and panaceas, and that history belongs in another volume. Now, we will turn to those publications that garnered interest and ideas in electricity as a power source.

During the nineteenth century, Britain witnessed a profound transformation not only in technological or material capability, but also in the cultural fabric of Victorian society. Electricity, perceived as scientific, entertaining, measurable, and useful, became quotidian, symbolizing progress, modernity, and the future. Public demonstrations of electrical phenomena captivated audiences and fostered a sense of wonder and optimism. In Britain, lectures at the Royal Institution became icons of education and progress, drawing crowds eager to witness the marvels of electricity. Writers like Mary Shelley in *Frankenstein* and H.G. Wells in *The Invisible Man* explored the potential and perils of electrical advancements, reflecting the embedding of these themes into Victorian culture.[27]

With the awareness and correspondence of Alessandro Volta's invention of the electric battery in 1800, electric power became a real possibility for electricians in and outside of the educated natural philosophers. The voltaic pile's ability to provide a continuous source of electrical current thus enabled subsequent experimental and practical applications in the field and in other burgeoning sciences. Experimenters like Humphry Davy harnessed it to delve deeper into electrochemistry and electrolysis, discovering numerous new elements and compounds. Davy's experiments, including isolating potassium and sodium, expanded the understanding of chemical processes. His work, in concert with natural philosophers around Europe and North America, laid the foundation for electrochemistry, influencing future scientific inquiries and industrial applications. Electrical energy became a tool for answering questions. Effectively, it became a commodity of inquiry.

In 1809, Davy created one of the first electric arc lamps, representing a significant milestone in the development of lighting. Davy passed an electric current through two carbon rods using a voltaic pile, producing a bright, sustained light arc. Although practical applications were limited at the time due to the unreliability of power supplies and the short lifespan of the carbon rods, this invention highlighted the immense potential of electric lighting outside of the laboratory and into places like streets, coal mines, and theaters. While not immediately practical or available, this invention showcased the potential for electric light to replace existing light sources. The electric arc lamp's brilliance and intensity were unparalleled at the time, and they attracted, repelled, and terrified viewers.

Arguably one of the most central figures in the history of electrical science, and certainly in Britain, Michael Faraday made seminal contributions to electromagnetism in the 1820s and 1830s, establishing the foundational principles of modern electrical power systems. Faraday's description of electromagnetic induction in 1831, which demonstrated that a changing magnetic field could induce an electric current in a conductor, is the fundamental principle behind electric generators and transformers – critical components of our current electrical power systems. Faraday's meticulous experiments, such as the creation of an electromagnetic generator, were deeply influenced by the earlier discoveries of Hans Christian Ørsted and André-Marie Ampère, which he built upon to revolutionize the field.[28] As an experimenter, correspondent, and believer in education, Faraday's impact extended beyond his discoveries because he helped popularize and demystify electricity for the general public.[29]

The invention of the telegraph, a dubious ream of discoveries and patents to untangle outside of this volume, did make possible the infrastructure through which people imagined distributing electric power. For example, the electric needle telegraph created by William Fothergill Cooke and Charles Wheatstone in 1837 revolutionized communication by significantly enhancing the speed and reliability of electrical currents transmitted over a great distance. Significantly, this system required fewer wires and utilized code that reduced telegraph staff needs. This achievement underscored the practical utility of electrical power for real-world applications and marked a pivotal moment in the history of telecommunications. Beyond his predictions, the stratifying of the world and cityscape with electric technologies literally and metaphorically remapped how people conceptualized time. The telegraph reduced communication times from days to minutes, profoundly affecting commerce, journalism, and personal communication. It facilitated the rapid dissemination of news and information, connecting distant parts of the country and the world and laying the groundwork for future developments in global communication networks. The telegraph's success demonstrated the immense potential of electrical systems, spurring further innovations and investments in electrical technologies.[30]

The development of electric motors in the 1830s and 1840s constituted another significant advancement. What qualifies as the first motor remains contested, but several inventors like William Sturgeon claimed the creation of the first practical electric motor in 1832, followed by subsequent improvements by inventors such as Thomas Davenport and English physicist Joseph Henry, showcased the potential for electric power in diverse industrial applications. These early motors converted electrical energy into mechanical motion, demonstrating the transformative potential of electric power. These innovations enabled the mechanization of various processes, leading to increased efficiency and productivity in manufacturing and other industries. The electric motor's development marked a significant step towards the modern industrial age, as it allowed for more precise and reliable machinery compared to steam-powered alternatives. This era of innovation laid the groundwork for future developments in transportation and manufacturing, revolutionizing the industrial landscape.[31]

As with any technoscientific breakthrough, electric power's success hinged upon peoples' trust and willingness to use it. Perhaps more than anything else, and certainly more than the erudite correspondences of members of the Royal Society, public demonstrations conducted the idea and imagination of electric power in Victorian culture.[32] Demonstrating electrical phenomena in the 1840s, often done by both prominent and less established scientists, played a crucial role in raising public awareness and interest in electricity. These demonstrations helped demystify the new technology and inspired further research and development in the field, fostering a broader understanding and acceptance of electrical innovations. These public lectures and demonstrations were educational and captivating, often drawing large crowds and receiving widespread media coverage. Maxwell's theoretical work on electromagnetism further solidified the scientific foundations laid by Faraday, contributing to the unification of electricity, magnetism, and optics into a single theory. Their efforts in public education and theoretical advancements made electricity an accessible and intriguing topic, encouraging a new generation of scientists and engineers to explore its possibilities.[33]

The 1850s marked the emergence of electrical engineering as a field of study and professional practice. Universities and technical institutions in England, such as King's College London, began offering courses and conducting research focused on the principles and applications of electricity. Influential figures such as William Thomson (later Lord Kelvin) contributed to the formalization of electrical engineering, which was essential for the systematic development and deployment of electrical technologies. His work on thermodynamics and electrical standards laid the groundwork for precise and reliable engineering practices. Establishing dedicated electrical engineering departments and publishing technical journals facilitated the exchange of knowledge and ideas, accelerating the field's growth. This period also saw the formation of professional organizations, which provided a platform for collaboration and the dissemination of best practices, further solidifying electrical engineering as a scientific discipline.[34]

The invention of the incandescent lamp in the 1860s was a pivotal development in the history of electric power. Although Thomas Edison is often credited with perfecting the incandescent light bulb in the 1870s, earlier contributions by English inventor Sir Joseph Swan were equally significant. Swan's work on developing a practical and long-lasting light bulb laid the groundwork for the widespread adoption of electric lighting. The rivalry and eventual collaboration between Swan and Edison further accelerated the development of electric lighting technology. Swan remarked, "Electric light will soon replace the gas light and illuminate our homes and streets." His pioneering efforts in improving the filament and vacuum within the bulb were crucial in making electric lighting viable for everyday use. This invention drastically changed living conditions, extending productive hours beyond daylight and improving public safety. The success of the incandescent lamp catalyzed further innovations in electrical appliances and systems, embedding electric lighting firmly into the fabric of modern life.[35]

The 1870s saw the initiation of efforts to develop systems for the widespread distribution of electrical power. Pioneers such as William Siemens and his brother Werner von Siemens worked on creating more efficient generators and distribution networks. Siemens' advancements in alternating current (AC) systems were particularly crucial, as they facilitated the transmission of electricity over longer distances with minimal loss, making centralized power generation and distribution feasible. These efforts included the development of the dynamo, which significantly improved the efficiency and practicality of power generation. The Siemens brothers' work laid the foundation for modern electrical grids, enabling the reliable and widespread use of electric power. Their contributions were instrumental in transitioning from localized, isolated power systems to interconnected networks capable of supporting growing urban and industrial demands.[36]

The first public electric lighting in England was installed in 1878 on the streets of Newcastle. This project, led by Joseph Swan, utilized his improved incandescent lamps and demonstrated the practicality and benefits of electric street lighting. The success of this project catalyzed the broader adoption of electric lighting in both public and private spaces, heralding the decline of gas lighting. Swan's partnership with the Edison & Swan United Electric Light Company significantly expanded the reach of electric lighting. Installing electric streetlights not only improved public safety by reducing crime rates but also enhanced urban life by making streets more navigable and businesses more attractive after dark. This early adoption in Newcastle served as a model for other cities, showcasing the transformative potential of electric lighting and encouraging further investments and innovations in the field.[37]

The 1880s witnessed the establishment of the first power stations in England, such as the Holborn Viaduct power station in London, which commenced operations in 1882. These early power stations, spearheaded by pioneers like Sir Coutts Lindsay and Sebastian Ziani de Ferranti, employed steam engines to drive electrical generators, providing power for street lighting and increasingly for residential and commercial use. The creation of power stations was a critical step in developing a reliable and scalable electric power infrastructure. The Holborn Viaduct station's success demonstrated the feasibility of centralized power generation and distribution, paving the way for more extensive and efficient electrical networks. These early power stations faced numerous technical and logistical challenges, but their eventual success underscored the viability and advantages of electric power, leading to rapid change and widespread adoption across the country.[38]

By the 1890s, the advantages of alternating current (AC) over direct current (DC) for power distribution had become widely recognized. AC systems, championed by engineers like Sebastian Ziani de Ferranti and John Hopkinson, enabled efficient transmission of electricity over long distances and facilitated the establishment of interconnected power grids. Ferranti's work on the Deptford Power Station in London exemplified the potential of large-scale AC power generation and distribution, demonstrating the feasibility and efficiency of this approach. The Deptford station, one of the first to use AC, highlighted the practicality of high-voltage transmission, reducing energy losses and making it possible to supply power to a wider area. This shift to AC systems marked a significant technological leap, addressing the limitations of DC systems and enabling the creation of more extensive and reliable electrical networks that could support growing urban and industrial demands.[39]

The electrification of industry in the late nineteenth century fundamentally transformed manufacturing and other sectors in England. Innovators such as Charles Parsons, who developed the steam turbine, and Elihu Thomson, who contributed to the development of electric welding, played crucial roles in this transformation. Electric motors and lighting enhanced productivity and working conditions in factories. Industries such as textiles, steel, and mining increasingly adopted electric power, resulting in greater efficiency and enabling the operation of machinery that was previously impractical with steam or manual power. The introduction of electric power not only increased production capacity but also improved safety and working conditions by reducing the hazards associated with steam power and manual labor. This widespread industrial electrification contributed to economic growth and positioned England at the forefront of technological innovation during the Industrial Revolution.[40]

The formation of numerous electrical companies in England during the late nineteenth century was instrumental in commercializing and expanding electrical technologies. Companies like the Edison & Swan United Electric Light Company, founded in 1883 by Joseph Swan and Thomas Edison, and the British Electric Traction Company, established in 1896, played crucial roles in developing and deploying electric lighting, trams, and other applications nationwide. Influential entrepreneurs such as William Preece, also contributed to establishing and growing these enterprises. These companies were pivotal in standardizing and mass-producing electrical equipment, making it more accessible and affordable for consumers and businesses. Their efforts in marketing and infrastructure development facilitated the rapid adoption of electric technologies, transforming urban and rural areas and integrating electricity into everyday life.[41]

By the end of the nineteenth century, an expansive engineering profession, consumer base, and utilities infrastructure fabricated a growing pressure for standardization and regulation. Establishing voltage, frequency, and safety protocol standards helped ensure compatibility and reliability across different systems and devices. Regulatory bodies and professional organizations, such as the Institution of Electrical Engineers (IEE), founded in 1871 by Carl Wilhelm Siemens, were pivotal in overseeing the development and implementation of these standards, fostering a more organized and efficient electrical industry in England. Figures like Oliver Lodge were influential in developing radio and wireless telegraphy, further contributing to the field's regulatory framework. These standardization efforts facilitated the safe and reliable use of electrical systems and promoted innovation by providing a consistent framework within which engineers and inventors could operate, ensuring that the benefits of electrical power could be broadly and effectively realized. By the end of the nineteenth century, the pervasive influence of electric power had firmly established itself in Victorian culture. The electrification of industries and urban centers fostered economic growth, while the development of power stations and electrical infrastructure paved the way for modern public works. Figures like Joseph Swan and Sebastian Ziani de Ferranti became emblematic of the era's inventive spirit, with their contributions fundamentally altering the landscape of British industry and daily life. As electricity became integral to Victorian society, it illuminated homes, streets, and the path toward a future driven by scientific advancement and technological progress.

ELECTRIC POWER IN THE HISTORY OF SCIENCE AND TECHNOLOGY

Since the 1980s, historians have significantly altered their approach to studying the history of electricity, moving beyond narratives of technological triumph to more nuanced analyses that incorporate social, cultural, economic, and environmental dimensions. One of the key changes has been an increased focus on the social and cultural impact of electricity. Scholars like David E. Nye, in his influential book *Electrifying America*, examine how electricity transformed social practices, domestic life, and urban development. Nye's work highlights how electric lighting changed nightlife, domestic routines, and cultural perceptions of modernity and progress, offering a richer understanding of how electricity influenced daily life. More recent scholarship, such as *Electrical Conquest,* edited by Bernard Carlson and Eric Conway, highlights how scholars have continued to expand the breadth of analysis.[42]

The economic and industrial context of electrification has also been scrutinized more deeply by historians, such as Thomas P. Hughes and Jonathan Coopersmith, who explore how electricity facilitated new industrial processes, altered labor practices, and contributed to the rise of new industries. Hughes' systemic analysis reveals the economic drivers and consequences of electrification, highlighting its impact on productivity and economic structures, thus providing a more comprehensive picture of how electricity powered industrial growth. For decades, historians have increasingly adopted the concept of technological systems to examine electrical infrastructure development.[43]

The environmental history of electricity has gained prominence as historians have examined how electric power generation and consumption impact the environment. This perspective includes the analysis of public policies and societal attitudes toward energy conservation and environmental protection, adding an essential dimension to the history of electricity. Scholars have examined the intricate spaces in which people embed electric power infrastructures and explore the interplay between energy, people, and their environment.[44] Historians have broadened their scope to include global and comparative electrification studies. Numerous works have developed comprehensive studies of electrification processes in different regions, revealing pathways' diversity and local conditions' influence. Such comparative studies illustrate how different countries and regions experienced and managed electrification, contributing to a more global understanding of technological change.[45]

The everyday experiences of electricity users have become a prominent area of focus in which research uncovers how ordinary people interacted with and adapted to electrical technologies in their homes and workplaces. They explore how electrical appliances changed domestic labor and how electrification affected rural versus urban areas, shedding light on the role of consumers in influencing the development of electrical goods and services. The intersection of electricity with gender and class has been another significant study area. This research reveals how access to and control over electrical resources varied across different social classes, highlighting disparities and the social implications of technological change.[46]

Historiography and methodology in studying electricity have evolved, with historians reflecting critically on earlier narratives and incorporating interdisciplinary methods. These changes in historical perspectives reflect broader trends in historiography toward more inclusive, interdisciplinary, and complex analyses of technological and societal developments. By considering a wider range of factors and voices, historians provide a richer and more nuanced understanding of the history of electricity and its impacts on Victorian society and its vast reach. This shift in focus enriches our understanding of the past and informs contemporary discussions on technology and its role in society.

VOLUMES AND CHRONOLOGY

This series of primary source volumes on *Electric Power in Victorian Britain* aims to explore the array of ideas, technologies, and conversations about electric power throughout Britain and across the nineteenth century, but we should acknowledge here, too, that they had a much wider influence as well. As I aim to show in these volumes, electrification is an incredibly broad process, and, perhaps more importantly, during the nineteenth century, it remained largely undefined. That is why the first volume, *Electric Power Imagined,* highlights how people fantasized and feared what electricity might do. It is only with this in mind that we can begin to appreciate the more concrete works on electric power physics and engineering later in the nineteenth century.

In the second volume, *Electric Power in Practice*, I turn to the many attempts to apply electricity to lighting, medicine, generating power, industry, and transportation. While many of the sources in this volume concern the invention of well-known technologies like the electric light or electromagnetic motor, it is essential to also note the technological and scientific ambiguity that surrounded the work of many of the scientists, engineers, and tinkerers involved. No set idea or certainty about the future of electric power existed. Indeed, many wished for an electrical future, but there remained opportunities for new imaginings. Electric power applications infiltrated all sorts of media incorporated the scientific elite, the public, and politicians.

In our third and final volume, *Electric Power, Politics and Culture*, I turn more overtly to the perceived political implications of electric power as applied to conversations about political power at home and in the British Empire. Electric power provides an avenue through which historians can explore how the British used electricity as a signifier of their technological superiority over other civilizations and races and also found it an incredible mechanism for restructuring government and control. The bureaucratic infrastructure and material networks that developed in Britain and its Empire played a crucial role in establishing the modes of electric power generation, transmission, and consumption that took shape at the close of the century. The remaining sources in this volume touch on some of the ways that electric power in Victorian culture, namely in literature, professionalization, and domestication. One of the most important historiographic shifts to take place in the history of electrification in the last few decades has been the reprioritization of users in narratives of electric power science and technology. How people choose to use, understand, or even identify with electric power must be a critical piece of how historians analyze electrification. People experienced electricity in laboratories, books, in school, and at public demonstrations. Any meaningful history of electrification must exploration of it must include these spaces.

Notes

1 Rhys Morus, I. (2015), "No Mere Dream: Material Culture and Electrical Imagination in Late Victorian Britain," *Centaurus*, 57: 173–191. https://doi.org/10.1111/1600–0498.12093

2 Ronald Wenzlhuemer, *Connecting the Nineteenth-Century World* (New York: Cambridge University Press, 2013).

3 Cantor, G. N., Dawson, G, Gooday G., Noakes, R., Shuttleworth, S., and Topham, J. *Science in the Nineteenth-Century Periodical: Reading the Magazine of Nature* (Cambridge University Press, 2004).

4 Hunt, Bruce, *Imperial Science: Cable Telegraphy and Electrical Physics in the Victorian British Empire* (Cambridge, 2021).

5 Kander, A., Malanima, P., and Warde, P., *Power to the People: Energy in Europe over the Last Five Centuries* (Princeton, 2013).

6 Cookson and Hempstead, *A Victorian Scientist and Engineer: Fleeming Jenkin and the Birth of Electrical Engineering* (Routledge, 2018).

7 Fleury Mottelay, *Bibliographical History of Electricity and Magnetism* (London: Charles Griffin and Co., 1922); Vaclav Smil, *Energy and Civilization: A History*

(Boston: The MIT Press, 2018); Richard Rhodes, *Energy: A Human History* (New York: Simon and Schuster, 2018).

8 Lianshan Chen, *Chinese Myths and Legends* (London: Cambridge University Press, 2011), 52.

9 Ken Dowden, *Zeus* (London: Taylor and Francis, 2006), 56.

10 Kedar, Yael, "'Virtus' and 'species' in the Philosophy of Nature of Roger Bacon (c. 1220–1293)" *Revista Española de Filosofía Medieval* 28 (2022): 43–65.

11 Pansy Hawk Wing, "Lakota Teachings: Inipi, Humleciya, and Yuwipi Ceremonies," in *Sacred Heritage* (Routledge, 1997).

12 King, D. N. T., Skipper, A., and Tawhai, W. B. "Māoei Environmental Knowledge of Local Weather and Climate Change," *Climate Change* 90 (2008): 385–409.

13 Gilbert, William, *De Magnete, Magneticisque Corporibus, et de Magno Tellure: Physiologia noua, plurimis &argumentis, & experimentis demonstrate* (London, 1600).

14 Kander, A., Malanima, P., and Warde, P., *Power to the People: Energy in Europe over the Last Five Centuries* (Princeton, 2013).

15 Sgarbi, Marco. *The Aristotelian Tradition and the Rise of British Empiricism: Logic and Epistemology in the British Isles, 1570–1689* (Springer, 2013).

16 Newton, Isaac., "Hypothesis Explaining the Properties of Light," in *The History of the Royal Society*, Thomas Birch ed. (London, 1757), 247–305.

17 Robert Boyle, "Experiments, notes, &c. about the mechanical origine or production of divers particular qualities among which is inferred a discourse of the imperfection of the chymist's doctrine of qualities: together with some reflections upon the hypothesis of alcali and acidum," https://name.umdl.umich.edu/A28980.0001.001. University of Michigan Library Digital Collections. Accessed October 22, 2024.

18 Brundtland, T. "From medicine to natural philosophy: Francis Hauksbee's way to the air-pump," T*he British Journal for the History of Science* 41 (2008): 209–240. https://doi.org/10.1017/S000708740800085X

19 Joseph Priestley, *The History and Present State of Electricity* (London, 1767), xv.

20 Letter from Stephen Gray to Dr. Mortimer "Some Experiments Relating to Electricity," *Philosophical Transactions*, December 31, 1735, 166–170.

21 Ian Jackson, "Science as Spectacle: Electrical Showmanship in the English Enlightenment," in *Frankenstein's Science* (Routledge, 2008).

22 Joseph Drury, *Novel Machines: Technolgy and Narrative Form in Enlightenment Britain* (Oxford, 2017).

23 Musschenbroek to Réaumur, January 20, 1746, *Procès-verbaux de l'Académie Royale des Sciences* 65, 6. Last Accessed May 5, 2024 https://gallica.bnf.fr/ark:/12148/bpt6k55741q/f2.item

24 Heilbron, 490.

25 Luigi Galvani, *De viribus electricitatis in motu musculari* (1792).

26 Alessandro Volta to Royal Society, March 20, 1800.

27 Koen Vermeir, "Electricity and Imagination: Post-romantic Electrified Experience and the Gendered Body: An Introduction," *Centaurus* 57. No. 3 (2015): 131–155.

28 Friedrich Steinle, trans. Alex Levine. *Exploratory Experiments: Ampère, Faraday, and the Origins of Electrodynamics* (University of Pittsburgh, 2016), 18–20.

29 Tiffany Watt-Smith, "Cardboard, Conjuring and "A Very Curious Experiment," *Interdisciplinary Science Reviews* 38 (2009): 306–320.

30 Arapostathis Stathis and Graeme Gooday, *Patently Contestable: Electrical Technologies and Inventor Identities on Trial in Britain* (MIT Press, 2013), 141.

31 Michael Bryan Schiffer, *Power Struggles: Scientific Authority and the Creation of Practical Electricity before Edison* (The MIT Press, 2008), 116–117.

32 Stephen Courtney, "'A Very Diadem of Light': Exhibitions in Victorian London, the Parliamentary Light and the Shaping of the Trinity House Lighthouses," *British Journal for the History of Science* 50, no. 2 (2017): 249–265.

33 Alan Hirshfeld, *The Electric Life of Michael Faraday* (Bloomsbury, 2009), xiii.
34 Crosbie Smith and Norton Wise, *Energy and Empire: Biographical Study of Lord Kelvin* (Cambridge, 1989), 483.
35 Chris Otter, *The Victorian Eye: A Political History of Light and Vision in Britain, 1800–1910* (Chicago, 2008), 13–14.
36 Wilfried Feldenkirchen, *Werner von Siemens: Inventor and International Entrepreneur* (Ohio State University, 1994), 84.
37 Leslie Hannah, *Electricity before Nationalization* (MacMillan, 1979), 28.
38 Thomas Hughes, *Networks of Power*, 55.
39 Bernard Carleson, *Tesla: The Inventor of the Electric Age* (Princeton, 2015), 89.
40 Richard Hill, *Power from Steam: A History of thew Stationary Steam Engine* (Cambridge, 1989), 231.
41 Kennedy WP, Delargy PJR. "Shorting the Future? Capital Markets and the Launch of the British Electrical Industry, 1882–1892, "*Business History Review*. 94, no. 2 (2020): 287–320.
42 David Nye, *Electrifying America: Social Meanings of a New Technology, 1880–1940* (MIT, 1992); Bernard Carlson and Erik Conway, *Electrical Conquest: New Approached to the History of Electrification* (Springer, 2023).
43 Thomas Hughes, *Networks of Power: Electrification in Western Society, 1880–1930* (Johns Hopkins, 1983); Jonathan Coopersmith, *The Electrification of Russia, 1880–1926* (Cornell, 1992); Vincent Lagendijk, *Electrifying Europe: The Power of Europe in the Construction of Electricity Networks* (Askant, 2008).
44 Richard White, *The Organic Machine: The Remaking of the Columbia River* (Hill and Wang, 1996); Robert F. Lifset, *Power on the Hudson: Storm King Mountain and the Emergence of Modern American Environmentalism* (Pittsburgh, 2014); Daniel Macfarlane, *Fixing Niagara Falls: Environment, Energy, and Engineers at the World's Most Famous Waterfall* (Chicago University Press, 2020).
45 William J. Hausman, Peter Hertner, and Mira Wilkins, *Global Electrification: Multinational Enterprise and International Finance in the History of Light and Power, 1878–2007* (Cambridge, 2018).
46 Graeme Gooday, *Domesticating Electricity: Technology, Uncertainty and Gender, 1880–1914* (Pittsburgh, 2008); Abigail Harrison Moore and R. W. Sandwell, eds., *In a New Light: Histories of Women and Energy* (McGill-Queen's University Press, 2021); Richard F. Hirsch, *Powering American Farms: The Overlooked Origins of Rural Electrification* (Johns Hopkins, 2022).

VOLUME I INTRODUCTION

During the nineteenth century, Victorians approached electricity with excitement and trepidation. Writers, scientists, and the broader public often communicated their anxiety about electricity through cautionary tales and warnings. As technological development made electricity more accessible and more visible in everyday life – through electric lighting, telegraphs, and early electrical motors – the risks associated with electrical shock and fire became all too real. Articles in scientific journals and popular periodicals alike emphasized the need for careful management of electricity, portraying it as both a tool of progress and a potentially dangerous force that could harm both individuals and society if not properly handled. Electricity, in this context, was imbued with a duality: it could be tamed and put to productive use, but only under the right conditions. Left unchecked, it could wreak havoc.

At the same time, Victorians imagined electricity as a nearly magical force that could revolutionize life in previously unimaginable ways. The burgeoning field of electrical science promised to unlock new dimensions of human experience, and electricity itself was often seen as a metaphor for the dynamism and unpredictability of the modern world. It represented the unseen forces that seemed to animate an industrial society – energy, speed, and communication – providing a fitting symbol for an era marked by the perception of constant motion and innovation. In fiction, authors utilized electricity as a means of invoking alchemical power capable of reanimating the dead, transforming bodies, or granting extraordinary abilities. They began to imagine what kind of transformations this powerful force might bring about in the real world.

Experiments with electricity were widespread during the nineteenth century, and many Victorians actively participated in the scientific exploration of its properties. Public demonstrations, lectures, and exhibitions allowed audiences to witness the wonders of electrical phenomena firsthand. The Royal Institution in London, under the guidance of people like Michael Faraday, became a hub for electrical experimentation, where Faraday's lectures on electricity captivated audiences with their displays of electrical sparks, magnetic fields, and mysterious arcs of light. Faraday's work on electromagnetism laid the foundation for the practical technology that would follow. Still, his experiments also had a profound cultural impact, elevating electricity from a subject of curiosity to one of serious scientific inquiry.

Faraday's experiments and those of his contemporaries helped popularize the idea of electricity as a fundamental force of nature that could be studied, quantified, and eventually harnessed for practical purposes. This shift in thinking reflected broader changes in Victorian scientific culture, where natural philosophers were increasingly seen as professionals rather than gentleman amateurs and scientific experimentation became more rigorous and methodical. Once seen as an unpredictable and dangerous curiosity, people increasingly perceived electricity as controllable through these experiments, its mysteries unpacked through the careful application of scientific principles.

The enthusiasm for electrical experimentation extended beyond the ranks of people considered professional scientists. "Amateur" scientists and inventors also conducted their own experiments with electricity, often with mixed results. Hobbyists built makeshift electrical devices in households across Britain, conducting experiments with Leyden jars, early batteries, and homemade electromagnets. These experiments were often dangerous – electric shocks were common, and fires could easily result from improperly managed experiments – but they fed the growing fascination with electricity as a force that could be mastered through ingenuity and persistence. Victorians' eagerness to experiment with electricity, despite its dangers, reflected their broader confidence in the possibilities of progress.

As electrical science transformed, so too did the technological applications of electricity. By the second half of the century, electric telegraphs had revolutionized communication, electric lights began to replace gas lamps in urban centers, and electrical motors powered new kinds of machinery. These practical uses of electricity transformed the daily lives of many Victorians, making electricity not just a scientific concept but a vital component of modern life. With the spread of electrification, electricity became synonymous with progress and modernity, a symbol of the Victorian belief in the power of technology to improve society. Even as electricity became more familiar, it retained an aura of mystery. The invisible forces that powered electric telegraphs and lit up city streets still defied easy comprehension, and many Victorians continued to imagine electricity as something almost supernatural. This tension between the known and the unknown, the practical and the fantastical, defined much of the Victorian engagement with electricity. It was at once a tool and a mystery, a source of anxiety and wonder.

The social implications of electricity were profound. In industrial settings, electricity offered new possibilities for production, efficiency, and control. Factories equipped with electric lighting could operate longer hours, and the development of electrical motors promised to reshape manufacturing processes. Electricity became associated with productivity and modernity, and its use in industry was seen as a marker of progress. However, the growing reliance on electricity also sparked fears about the dehumanizing effects of industrialization. Just as steam power had transformed labor in the previous century, electrical power now seemed poised further to alienate workers from the products of their labor, fueling anxieties about the relationship between technology and society.

By the end of the nineteenth century, electricity had become a defining feature of Victorian culture. It was woven into the fabric of everyday life, from the lighting of homes and streets to the powering of factories and the transmission of information. Electricity represented both the promise and the perils of the modern age, embodying the Victorian belief in the power of science to unlock the mysteries of nature and transform the world. At the same time, it served as a reminder of the limits of human knowledge and control, a force that could never be fully tamed or understood. In approaching electricity, Victorians grappled with a complex and evolving set of questions. What was this mysterious force, and what could it do? How could it be harnessed safely and effectively, and what dangers did it pose? As they experimented with electricity and explored its possibilities, they helped lay the groundwork for the electrified world of the twentieth century, even as they remained haunted by the questions that electricity raised.

Part 1

ANXIETY

Editorial Headnote

1. *Reflections on the Decline of Science in England, and on some of its Causes* (London: 1830), pp. 14–17.

Charles Babbage's *Reflections on the Decline of Science in England* (1830) is a critical examination of the state of scientific inquiry and institutional support in early nineteenth-century England. In the specified section, Babbage articulates his concerns about the factors contributing to the perceived decline in scientific progress and offers recommendations for revitalizing the field. He begins by lamenting the lack of adequate support for scientific research in England, contrasting it with the more robust systems in other European countries. He highlights the insufficient funding and resources allocated to scientific endeavors, pointing out that many talented individuals cannot pursue their research due to financial constraints. Babbage argues that this neglect stifles innovation and undermines England's potential to contribute to global scientific advancements.

A significant portion of Babbage's critique focuses on the inefficiencies and mismanagement within scientific institutions. He identifies a lack of accountability and transparency in the administration of these bodies, which he believes leads to the misallocation of resources and favoritism. Babbage calls for reforms to ensure that these institutions are led by individuals with genuine scientific expertise and a commitment to advancing knowledge, rather than by those with political or social connections. Babbage also addresses the cultural attitudes towards science in England, noting a general apathy and lack of appreciation for scientific achievements. He argues that this cultural milieu discourages young people from pursuing careers in science and leads to a dearth of new talent in the field. Babbage advocates for a greater emphasis on science education at all levels, from primary schools to universities, to foster a new generation of scientifically literate and enthusiastic individuals.

Moreover, Babbage critiques the dissemination of scientific knowledge, pointing out that many scientific findings are not effectively communicated to the broader public or even within the scientific community. He calls for improvements in the publication and distribution of scientific work, suggesting that better access to information would stimulate collaboration and innovation.

In his reflections, Babbage emphasizes the importance of international collaboration and the exchange of ideas. He believes that by fostering stronger connections with scientists from other countries, England can benefit from diverse perspectives and stay at the forefront of scientific discovery. Babbage envisions a more interconnected and supportive global scientific community where knowledge and resources are shared for the collective advancement of science.

In summary, Charles Babbage's *Reflections on the Decline of Science in England* provides a critical analysis of the challenges facing English science in the early nineteenth century. Through his examination of institutional inefficiencies, cultural attitudes, and communication barriers, Babbage offers a comprehensive critique and a call to action for revitalizing scientific inquiry and fostering a more supportive environment for innovation and discovery.

1

REFLECTIONS ON THE DECLINE OF SCIENCE IN ENGLAND, AND ON SOME OF ITS CAUSES

Charles Babbage

Source: Charles Babbage, *Reflections on the Decline of Science in England, and on some of its Causes* (London: 1830), pp. 14–17.

OF NATIONAL ENCOURAGEMENT

The little encouragement which at all previous periods has been afforded by the English Government to the authors of useful discoveries, or of new and valuable inventions, is justified on the following grounds:

1. The public, who consume the new commodity or profit by the new invention, are much better judges of its merit than the government can be.
2. The reward which arises from the sale of the commodity is usually much larger than that which government would be justified in bestowing; and it is exactly proportioned to the consumption, that is, to the want which the public feel for the new article.

It must be admitted that, as general principles, these are correct: there are, however, exceptions which flow necessarily from the very reasoning from which they were deduced. Without entering minutely into these exceptions, it will be sufficient to show that all abstract truth is entirely excluded from reward under this system. It is only the application of principles to common life which can be thus rewarded. A few instances may perhaps render this position more evident. The principle of the hydrostatic paradox was known as a speculative truth in the time of Stevinus[1]; and its application to raising heavy weights has long been stated in elementary treatises on natural philosophy, as well as constantly exhibited in lectures. Yet, it may fairly be regarded as a mere abstract principle, until the late Mr. Bramah, by substituting a pump instead of the smaller column, converted it into a most valuable and powerful engine. – The principle of the convertibility of the centres of oscillation and suspension in the pendulum, discovered by Huygens more than a century and a half ago, remained, until within these few years,

DOI: 10.4324/9781003295556-2

a sterile, though most elegant proposition; when, after being hinted at by Prony, and distinctly pointed out by Bonenberger, it was employed by Captain Kater as the foundation of a most convenient practical method of determining the length of the pendulum. – The interval which separated the discovery, by Dr. Black, of latent heat, from the beautiful and successful, application of it to the steam engine, was comparatively short; but it required the efforts of two minds; and both were of the highest order. – The influence of electricity in producing decompositions, although of inestimable value as an instrument of discovery in chemical inquiries, can hardly be said to have been applied to the practical purposes of life, until the same powerful genius[2] which detected the principle, applied it, by a singular felicity of reasoning, to arrest the corrosion of the copper-sheathing of vessels. That admirably connected chain of reasoning, the truth of which is confirmed by its very failure as a remedy,[3] will probably at some future day supply, by its successful application, a new proof of the position we are endeavouring to establish.

Other instances might, if necessary, be adduced, to show that long intervals frequently elapse between the discovery of new principles in science and their practical application: nor ought this at all to surprise us. Those intellectual qualifications, which give birth to new principles or to new methods, are of quite a different order from those which are necessary for their practical application.

Source Endnotes

1 About the year 1600.

2 I am authorised in stating, that this was regarded by Laplace as the greatest of Sir Humphry Davy's discoveries.

3 It did *not* fail in producing the effect foreseen by Sir H. Davy, – *the preventing the corrosion of the copper;* but it failed as a cure of the evil, by producing one of an *opposite* character; either by preserving too perfectly from decay the surface of the copper, or by rendering it negative, it allowed marine animals and vegetables to accumulate on its surface, and thus impede the progress of the vessel.

Editorial Headnote

2. "What Will He Grow To?", *Punch*, June 25, 1881.

Few sources successfully expose the excitement, anxiety, and significance of Victorian electrification like the included comic in *Punch*. Published in 1881, as electric power began pulsing the waterwheel mill in Godalming, Surrey, the comic powerfully exposes the underdetermined future of electric power. Punch famously satirized the country's politics, business, and scientists throughout the nineteenth century and during the 1870s and 1880s, featured numerous articles and comics about the hazards, benefits, and absurdities of electric lighting, power, medicine, spiritualism, and telegraphy. For historians of electric power systems, it is essential to wrestle with this moment of uncertainty. Not only did the applications and possibilities of electricity captivate the experimenters, tinkerers, and physicists studying electrical phenomena, but it also kept the attention of the public. Perhaps electric power would replace the existing energy titans or just become another one of them. Or perhaps electric power, and specifically the storage of force, could change the landscape of industrialization by redistributing the control of resources, clearing the smog, or even removing Great Britain's supremacy in manufacturing. Although electric power systems in Victorian Britain did not really do any of those things, and actually often simply intensified the problems people had with earlier power systems, this comic works as a powerful reminder that they did not have to. People chose what electric power systems did, who they impacted, and how they reshaped the world.

2

WHAT WILL HE GROW TO? (ILLUSTRATION)

Source: "What Will He Grow To?", *Punch*, June 25, 1881.

PUNCH, OR THE LONDON CHARIVARI.—JUNE 25, 1881.

"WHAT WILL HE GROW TO?"

DOI: 10.4324/9781003295556-3

Editorial Headnote

3. *The Coal Question: An Inquiry Concerning the Progress of the Nation, and the Probable Exhaustion of Our Coal-Mines* (London: MacMillan and Co., 1865), pp. 138–145.

In *The Coal Question* (1865), economist William Stanley Jevons examines Britain's dependence on coal as a finite resource, presenting coal depletion as a pressing threat to national prosperity and industrial strength. In Chapter 7, "Of the Supposed Substitutes for Coal," Jevons notably considers electricity as a potential alternative to coal, which he terms "the mainspring of our progress." He posits electricity as a future-oriented substitute that might alleviate Britain's energy concerns, but he remains cautious, suggesting that electricity's limitations as a direct replacement still necessitate prudent coal usage. By framing electricity as a substitute, Jevons illustrates early anxieties about resource scarcity in an expanding empire, where Britain's access to energy – primarily through coal – was both a measure of its global dominance and a vulnerability.

Jevons's treatment of electricity as a possible replacement for coal is significant because it reflects growing imperial concerns about maintaining industrial and military superiority as Britain approached what he believed was an inevitable energy crisis. The imperial implications of his analysis are twofold. First, the suggestion of electricity as an alternative energy source underscores British efforts to sustain dominance by exploring innovations that might secure continued industrial production without coal reliance. Second, his reservations about electricity's current limitations highlight imperial anxieties about energy security; without sufficient coal reserves or reliable substitutes, Britain's infrastructure and, by extension, its empire would face existential risks. Jevons thus taps into a broader Victorian anxiety about the durability of Britain's imperial economy, implicitly linking domestic energy policy with international power dynamics. His work prefigures later debates on the limits of industrial expansion and reveals an early awareness of the environmental and geopolitical strains underlying imperial ambitions.

3

THE COAL QUESTION

An Inquiry Concerning the Progress of the Nation, and the Probable Exhaustion of Our Coal-Mines

William Stanley Jevons

Source: William Stanley Jevons, *The Coal Question: An Inquiry Concerning the Progress of the Nation, and the Probable Exhaustion of Our Coal-Mines* (London: MacMillan and Co., 1865), pp. 138–145.

Hitherto we have considered mechanical force only, but it is obvious that if coal were used up we should want some source of heat as well as force. A favourite notion is to employ wind, water, or tidal mills to turn magneto-electric machines, and by the stream of electricity produced to decompose water, thus furnishing a continuous supply of artificial gaseous fuel. Such a plan was proposed in the *Times* during the discussions on the French Treaty. But an answer, attributed to Dr. Percy of the School of Mines, soon appeared, showing the amount of fuel derivable to be inconsiderable. The waste of power must be vastly greater in such a process of transmutation than in the system of artificial water power which we have considered. Besides, if uniform experience is to be trusted, a steam-engine would be a much more economical means of turning the magneto-electric machines than either a wind, water, or tidal machine. We should therefore only use coal in a roundabout manner to generate a less valuable fuel. For the hydrogen gas generated, though in some instances valuable, would in general be immensely less convenient than coal. For equal weights, it gives about four times as much heat as coal, but hydrogen is so light that *for equal volumes gives only one five thousandth part as much heat.* To compress it in a small space would require more force than the combustion of the fuel itself would furnish, and gas companies do not find it convenient to compress their gas. Hydrogen too has so much higher a diffusive power than coal-gas, that it could hardly be retained in gasometers or ordinary pipes. Even the loss of coal-gas by leakage is said to be nearly twenty-five per cent.

Of course it is useless to think of substituting any other kind of fuel for coal We cannot revert to timber fuel, for "nearly the entire surface of our island would be required to grow timber sufficient for the consumption of the iron manufacture alone."[1] And I have independently calculated, from the known produce of Continental forests,[2]

DOI: 10.4324/9781003295556-4

and the comparative heat-producing values of timber and coal,[3] that forests of an extent two and a half times exceeding the whole area of the United Kingdom would be required to furnish even a theoretical equivalent to our annual coal produce. Practically, however, there are inconveniences about the use of timber that would altogether prevent it from nourishing a large manufacturing system. Wood fuel is superior to coal in the single case of the iron smelting furnace, but in most other uses, the greater bulk of wood, and the large areas of forest land over which it is spread, necessarily render it a costly and inefficient fuel compared with coal.

Peat, or turf, again, may no doubt be turned into fuel; but, in spite of what has been said in its favour by Sir R. Kane,[4] all experience shows that it is immensely inferior as regards cost and efficiency to coal. It is usually full, too, of phosphorus and sulphur, and thus has not even those advantages of purity which render timber so valuable for the iron blast furnace.

Petroleum has of late years become the matter of a most extensive trade, and has even been proposed by American inventors for use in marine steam-engine boilers. It is undoubtedly superior to coal for many purposes, and is capable of replacing it. But then, *What is Petroleum hut the Essence of Coal,* distilled from it by terrestrial or artificial heat? Its natural supply is far more limited and uncertain than that of coal, its price is about 15*l*. per ton already, and an artificial supply can only be had by the distillation of some kind of coal at considerable cost. To extend the use of petroleum, then, is only a new way of pushing the consumption of coal. It is more likely to be an aggravation of the drain than a remedy.

Coal has all those characteristics which entitle it to be considered the best natural source of motive power. It is like a spring, wound up during geological ages for us to let down. Just as in alluvial deposits of gold-dust we enjoy the labour of the natural forces which for ages were breaking down the quartz veins and washing out the gold ready for us, so in our seams we have peculiar stores of force collected from the sunbeams for us. Coal contains light and heat bottled up in the earth, as Stephenson said, for tens of thousands of years, and now again brought forth and made to work for human purposes.

The amount of power contained in coal is almost incredible. In burning a single pound of coal there is force developed equivalent to that of 11,422,000 pounds weight falling one foot, and the actual useful force got from each pound of coal in a good steam-engine is that of 1,000,000 lbs. falling through a foot; that is to say, there is *spring* enough in coal to raise a million times its own weight a foot high. Or again, suppose a farmer to despatch a horse and cart to bring a ton of coals to work a portable engine, occupying four hours on the way. The power brought in the coal is 2,800 times the power expended in bringing it, and the amount of useful force actually got from it will probably exceed by 100 times or more that of the horse as employed in the cart. In coal we pre-eminently have, as the partner of Watt said, "What all the world wants, – POWER." All things considered, it is not reasonable to suppose or expect that the power of coal will ever be superseded by anything better. It is the naturally best source of power, as air and water and gold and iron are, each for its own purposes, the most useful of substances, and such as will never be superseded.

Of course I do not deny that if our coal were gone, or nearly so, and of high price, we might find wind, water, or tidal mills a profitable substitute for coal. But this would only be on the principle that half a loaf is better than no bread. It would not enable us to keep up our old efficiency, nor to compete with nations enjoying yet undiminished stores of fuel. And there is little doubt, too, that a century hence the steam-engine will be two or threefold as efficient as at present, turning the balance of economy so far the more in favour of those who then possess coal, and against those who have to resort to water or wind.

And, indeed, this is a point which I must insist upon as finally decisive of the question. *progress of science, and the improvements in the arts, will tend to increase the supremacy of steam and coal.* Any mechanist knows that the water-wheel and the windmill have been brought, by the exertion of our engineers, Brindley, Smeaton, Rennie, Telford and Fairbairn, near to their mathematical limit of efficiency; so that we can do little more than improve the mechanical construction, and gain some small per centage of additional power by reducing the friction of the machinery. The steam-engine, on the other hand, at least equally admits of improvement in mechanical details; but beyond this, in the principles of heat and vapour, we see clearly the possibility of multiplying at least three-fold the efficiency of fuel If there is anything certain in the progress of the arts and sciences it is that this gain of power will be achieved, and that all competition with the power of coal will then be out of the question. In short, the general course of science and improvement will only lead us the more to regret the limited extent of our coal resources.

But let us further remember that coal is now a pre-eminent gift in our actual possession, whereas if any wholly new source of power be some day discovered, we have no reason to suppose that our island will be as pre-eminently endowed with it as with coal.

Among the residual possibilities of unforeseen events, it is just possible that some day the sunbeams may be collected, or that some source of force now unknown may be detected. But such a discovery would simply destroy our peculiar industrial supremacy. Electricity has already been zealously cultivated on the Continent with this view, – "England," it is said, "is to lose her superiority as a manufacturing country, inasmuch as her vast store of coals will no longer avail her, as an economical source of motive power."[5] And while foreigners clearly see that the peculiar material energy of England depends on coal, we must not dwell in such a fool's paradise as to imagine we can do without coal what we do with it.

Source Endnotes

1 Taylor's Archaeology of the Coal Trade, p. 176.
2 Percy's Metallurgy, vol. i. pp. 71, 72.
3 Watt's Chemical Dictionary, Article, *Fuel.*
4 Industrial Resources of Ireland, 1st ed. chap. ii.
5 Liebig's Letters on Chemistry, No. 12, p. 154.

Part 2

IMAGINING ELECTRICITY

Editorial Headnote

4. "Historical Electricity," from *An Essay on the History, Practice, and Theory, of Electricity* (London: J. Johnson and Co., 1810), pp. 1–22.

Little is known of the British writer John Bywater, whose biography and academic record are mostly lost to history. He published several essays on experimentation, electrical science, magnetism, and compasses.[1] The introduction and text presented here, first published in 1810, offers an excellent window into the mind of what Bywater refers to as the "practical electrician," which was the academics, natural philosophers, and intellectual public consumed by all things electric who were the precursors to the tinkers, engineers, and electricians described by people like Bruce Hunt, Graeme Gooday, and Iwan Morus.[2] Bywater's "practical electricians" circulated throughout the academic establishment at universities like Glasgow and in the halls of British institutions like the Royal Society and Royal Institution. In these pages, Bywater highlighted key changes in what would be in the ensuing decades come to be termed electrical science in the nineteenth century. For historians of this period, the account Bywater provides illuminates the methods of experimentation circulating within the British press at the turn of the nineteenth century, displays the mythology people built around people like Benjamin Franklin (1706–1790) and Joseph Priestly (1733–1804), and highlights the significance with which some people inscribed electrical experiments within natural philosophical discourse. Electrical experiments signaled broader shifts in and celebrations about experimental science. Such writings on the history of electrical experimentation represent the growing appetite for increased demonstration of electrical energy's complexity and potential applications.

Editorial Endnotes

1 John Bywater. *Observations on the Deviation of the Compass* (London: Longman, 1820).

2 Bruce Hunt, *The Maxwellians* (Ithica: Cornell University press, 1991); Iwan Morus, *Frankenstein's Children: Electricity, Exhibition, and Experiment in Early-Nineteenth Century London* (Princeton: Princeton University Press, 1998); Graeme Gooday, *The Morals of Measurement: Accuracy, Irony, and Trust in Late Victorian Electrical Practice* (Cambridge: Cambridge Univrsity Press, 2004).

4

HISTORICAL ELECTRICITY

John Bywater

Source: John Bywater, "Historical Electricity," from *An Essay on the History, Practice, and Theory, of Electricity* (London: J. Johnson and Co., 1810), pp. 1–22.

PART. I.

FEW mental employments give more pleasure to the philosophic mind than tracing the history of the sciences, and marking out those peculiar and important events which have led to their present state of improvement.

By following the steps which led to many discoveries in electricity, we shall perceive the power and sagacity of genius, and that display of human intellect which cannot fail to raise our admiration: while, on the other hand, we shall find numerous and important discoveries have resulted from the unwearied researches of men who were not distinguished for their genius; and this consideration holds out the fairest prospect that the labours of the experimentalist will not go unrewarded.

When regarding the progressive advancement of knowledge, and the regular extension of the powers of the human mind, produced by the acquisitions of science, it is painful to observe how, in dark and superstitious ages, obstacles were thrown in the way of the studious, and attempts were made to excite a prejudice against the votaries of natural philosophy, by the most disgraceful perversion of their peaceful labours, to gratify an intolerant spirit of bigotry. Happily the rise and progress of electricity have not been clouded or perverted by the ignorance of early times, as its most important events have chiefly resulted from the inquiries of electricians who lived in the last century: the history of this science is therefore clear and instructive, and powerfully stimulates the mind to further investigation.

We are informed that the attractive power of amber was known several hundred years before the Christian era; yet it does not appear that the subject commanded the attention of the learned until the year 1600, when Dr. Gilbert, of Colchester, published a list of different substances that acquire this attractive property by a slight degree of friction. "This electrical attraction Dr. Gilbert supposed was performed in the same manner as the attraction of cohesion: two drops of water, he observed, rushed together when they are brought into contact; and electrics, he says, are virtually brought into contact with the bodies they act upon, by means of their effluvia, excited by friction." Mr. Boyle, who flourished about the year 1670,

DOI: 10.4324/9781003295556-6

afterwards added to the list of substances that Gilbert had published, and made some remarks on the electric light of diamonds.

After Gilbert and Boyle, Mr. Hawksbee, who wrote in 1709, distinguished himself by several new discoveries in electricity. This gentleman was the first who observed the great electric power of rubbed glass, the light and noise proceeding from it, together with a variety of phenomena relating to electrical attraction and repulsion. Some of the experiments made by Mr. Hawksbee, were so singular that they still deserve to be noticed. This gentleman "lined more than one half of the inside of a glass globe with sealing-wax, and, having exhausted the globe, he put it into motion; when applying his hand to excite it, he saw the shape and figure of all the parts of his hand, distinctly and perfectly, on the concave superficies of the wax within: it was as if there had only been pure glass, and no wax interposed between his eye and his hand. The lining of wax, where it was spread the thinnest, would but just allow the light of a candle to pass through it; and in some places the wax was at least one eighth of an inch thick, yet even in those places the light and figure of his hand were as distinguishable as in any other part." Mr. Hawksbee was not inattentive to the sound made by the emission of the electric matter, and observed that all the electric phenomena were improved by warmth, and diminished by moisture, which he proved by several experiments.

We next come to the labours of an experimentalist, perhaps more assiduous in this new field of inquiry than any of his predecessors. Mr. Stephen Grey, a pensioner at the Charter House, was the first that conveyed the electric fluid along certain bodies to any considerable distance. This he put into practice in the beginning of the year 1729, by exciting or rubbing a glass tube, while he stood in his balcony, and then bringing the excited part of the tube in contact with the upper part of an hempen string, which descended to the court below, for he found that a small ivory ball, which was attached to the lower end of the string, would attract and repel light bodies, when he had electrified the string with his excited glass tube. Mr. Grey next made an attempt to convey the electric fluid in an horizontal direction, and from his endeavours to accomplish this object arose a discovery of which he had not the least conception. To complete the object he had in view, he made a loop at each end of a packthread, and hung one end upon a nail driven into a beam; then, through the loop at the lower end of the packthread, he passed one end of the hempen line, along which the fluid was to be conveyed, and fastened the other end of the hempen line to his glass tube: by this means the greater part of the hempen line was suspended between his tube and the lower loop of the packthread, and lay in an horizontal direction, while another part of the line hung in a perpendicular direction from the packthread loop; and to the lower end of this part of the line was attached an ivory ball. After this preparation, Mr. G. expected that when small bodies were placed near the ivory ball, they would be attracted by the ball at the time the other end of the line was electrified by the tube; but to his very great surprise he could not convey the least electric fire to the ball. Soon after this he paid a visit to Mr. Wheeler, and told him of his fruitless attempt to convey the electric fluid along an hempen line suspended in an horizontal direction. Mr.

Wheeler then proposed to suspend the line to be electrified by another of silk, instead of packthread; Mr. Grey observed that it might do better on account of its smallness, as the electric fluid most probably would not pass so readily to the beam by this silk line as by the packthread. These gentlemen then suspended the linc to be electrified by silk, and found the experiment succeed far beyond their expectation.

We are now on the eve of a discovery that may justly be numbered among the most important discoveries in electricity: the following circumstances will however shew, that the discoverers owed more to their unremitting assiduity than to any great depth of reasoning. Elated with the success of their last experiment, these gentlemen afterwards tried to convey the electric fluid along a much longer hempen line, still supposing that their success depended on the smallness of the line that suspended the one along which the fluid was to be conveyed; this persuasion gave them considerable trouble, though in the end it led them to the important discovery to which I allude. After the communicating line had been much extended, its encreased weight, and the motion given it by rubbing the tube, broke the silk line by which it was suspended. This led them to imagine, that metal wire would answer their purpose better than silk, because wire of the same diameter as the silk line, would be much stronger: accordingly they suspended the communicating line by a brass wire, and electrified it with the glass tube, but, to their great astonishment they could not convey the least electric fluid to the ivory ball at the other end of the line; therefore, they were almost compelled to come to this conclusion, that the success of the experiment did not depend on the size of the lines, but on the nature of the materials of which those lines were constructed. This discovery gave birth to numberless new experiments: it was soon found that many other substances, besides silk, would not convey the electric fluid from one place to another: glass, resin, jet, copal, sealing wax, air, most dried animal substances, and all the precious stones, were of this kind; therefore all those bodies which would not convey the electric fluid were called electrics, or non-conductors of electricity; and all those bodies, such as stony and metallic substances, that would convey the electric fluid, were called conductors or non-electrics, and this distinction is retained to the present period.

After this time we find many inanimate, and also animate bodies, were suspended by silk lines, or placed on glass, resin, &c. to be filled with the electric fire. Mr. Grey suspended a pole, twenty-seven feet long, by hair lines, in an horizontal direction, and was enabled to convey the electric fluid from one end to the other. This gentleman then suspended a boy on hair lines, in a similar position, and found that the boy as well as the pole, was capable of being filled with electricity. "Mr. Grey having discovered that he could communicate electricity to a bottle of soap and water, was encouraged to attempt communicating it to the latter. In order to this, he electrified a wooden dish full of water, placed on a cake of rosin, or a pane of glass, and observed that if a small piece of thread, a narrow slip of thin paper, or a piece of sheet glass, was held over the water in an horizontal position, at the distance of an inch, or something more, they would be attracted to the surface of the water and then repelled; but he imagined that these attractions and repulsions were not repeated so often as they would have been if the body had been solid."

Mr. Du Fay, by pursuing Mr. Grey's experiments to make water receive electricity, found "that all bodies, without exception, whether solid or fluid, were capable of receiving it, when they were placed on glass or sealing-wax slightly warmed, or only dried, and the excited tube was brought near them; he particularly mentions his having made the experiment with ice, lighted wood, red hot iron, coal, and every thing that happened to be at hand at the time, and, to his great surprize, remarked that such bodies as were of themselves the least electric, had the greatest degree of electricity communicated to them, by the approach of the excited tube." Mr. Du Fay was the first that observed an electric spark from a living body; this gentleman having suspended himself on silk lines, "as Mr. Grey had done the boy mentioned above, observed, that as soon as he was electrified, if another person approached him, and brought his hand within an inch, or thereabouts, of his face, legs, hands, or cloaths, there immediately issued from his body one of more pricking shoots, attended with a crackling noise; he says this experiment occasioned to the person who brought his hand near him, as well as to himself, a little pain resembling that of a sudden prick of a pin, or the burning from a spark of fire; and that it was felt as sensibly through his cloaths, as on his bare face or hands. He also observes, that, in the dark, those snappings were so many sparks of fire."

Mr. Grey and his friends suspended an iron-rod, four feet long, and half an inch in diameter, upon silk lines, and by applying the excited tube to one end of it, they perceived not only a light upon that end of the rod to which they applied the tube, but a stream of electric fire issuing from the other end at the same time; they then connected a pewter plate with the iron rod, and found the plate produced similar effects to the iron rod when electrified. They afterwards filled the plate with water, and Mr. Grey observes, "they found the same light and the same pricking at the finger, and the same snapping, as when the experiment was made on the empty plate." These experiments appeared so singular, that they induced a general spirit for electrical inquiry, in several parts of Europe, and in which Germany took a considerable share.

About 1742, Mr. Boze, a German philosopher, substituted a glass globe instead of the glass tube, and, by rubbing it with the hand while it turned on its axis, he was enabled to excite a much greater quantity of electric fire than could be obtained by the glass tube. This gentleman was the first who suspended an iron or tin tube on silk lines, to receive the electric fire from the excited globe; and to prevent the tube from injuring the glass, he hung a bundle of threads from one end of the tube, which reached to the surface of the globe, in order to collect the electric matter.

Mr. Winckler, of Leipsic, substituted a cushion instead of the hand; and Mr. Gordon then adopted a cylinder instead of a globe, which is generally adopted at the present period. The accumulation of electricity, acquired by these means, gave a more important air to electrical experiments, and imparted fresh energy to the lovers of this delightful study.

In England, Dr. Watson took the lead, and furnished the public with many entertaining experiments. This gentleman fired spirits of wine, inflammable air, and gunpowder, by the electric spark. He afterwards imitated the Germans, and excited four glass globes at the same time; and found, when "two pewter plates were held, one in the hand of an electrified person, and the other in the hand of one who stood upon the floor, the flashes of pure and bright flame were so large, and succeeded each other so fast, that, when the room was darkened, he could distinctly see the faces of thirteen persons, who stood round the room."

In the preceding observations, it has been my intention to arrange the most important discoveries in such a manner, as to enable those who know but little of electricity, to follow, step by step, the advances of this science, and, in some measure, to anticipate the experiment that led to a discovery whose effects are the most singular in the history of natural philosophy. We have seen that the discoveries of Messrs. Grey and Wheeler enabled electricians to charge all conducting bodies with electricity, when those bodies were suspended by silk lines, or placed on any non-conducting body; therefore, we shall perceive, that this circumstance naturally suggested the experiment that led to the discovery to which I allude.

Professor Muschenbroeck observed, that when conducting bodies were placed on glass, &c. and electrified, that their electricity was very soon carried off by the conducting particles floating in the atmosphere; he therefore imagined, that if a conducting substance was put into a glass phial, that it could be charged much higher than in open air, as the glass would protect it from the dissipating action of the atmosphere. About 1745, he tried to put this idea in practice, by filling a small phial with water, which is a conducting substance. In making this experiment, the professor passed the end of a wire through the cork of the phial, so as to touch the water, and then charged the water by bringing the wire in contact with the prime conductor, but found no extraordinary result from the experiment. Mr. Cuneus, of Leyden, who was one of the party when the professor made the experiment, repeated it afterwards; and happening to hold the phial in his hand, after he had connected the wire with the prime conductor, until the water, as he supposed, had received a full charge of electricity, and then applying his other hand to unloose the wire from the conductor, he received such a sudden shock in his arms and breast, as filled him with astonishment.

Thus we see how gradually these philosophers were led to the discovery of what is called the Leyden experiment, an event on which the eyes of Europe were immediately turned with eager attention. Dr. Priestley remarks, "It was this experiment that gave eclat to electricity, from this time it became the subject of general conversation, and every one was eager to see and feel the experiment, though such terrible accounts had been given of its effects. The most remarkable account is that of Mr. Winckler of Leipsic; he says, that the first time he tried the Leyden experiment, he found great convulsions by it in his body; and that it put his blood into great agitation; so that he was afraid of an ardent fever, and was obliged to use refrigerating medicines; he also felt a heaviness in his head, as if a stone lay upon it. Twice, he says, it gave him a bleeding at the nose, to which he was not inclined. His wife (whose curiosity it seems was stronger than her fears) received the shock only twice, and found herself so weak, that she could hardly walk; and a week after, upon recovering courage to receive another shock, she bled at the nose after taking it only once."

Professor Muschenbroeck tried the experiment with a thin glass bowl, and said, that he would not take a second for the kingdom of France. "But we are not to infer" (Dr. Priestley remarks) "from these instances, that all electricians were struck with this panic. Few, I believe, would have joined with the cowardly professor, who said, that he would not take a second for the kingdom of France. Far different from these were the sentiments of the magnanimous Mr. Boze, who, with a truly philosophical heroism, worthy of the renowned Empedocles, said, he wished he might die by the electric shock, and furnish an article for the Memoirs of the French Academy of Sciences. But it is not given to every electrician to die in so glorious a manner as the justly envied Richman."[1]

The Leyden experiments, made about 1746, were chiefly made with phials filled with water. But in 1747, Dr. Bevis covered a pane of glass on both sides with tin-foil, excepting about an inch from the edge, all round the pane, and found that it gave very powerful shocks. Afterwards Dr. Watson coated large jars, both inside and out, with leaf silver, and the power of the Leyden experiment became encreased by these means to such a degree, as to melt wires, and kill birds and fishes. In the next advances of this science we shall have to consider the effects of the Leyden experiment on a more extended scale of action. The French philosophers were the first to appear in this new field of inquiry: they made a shock pass through wire two thousand toises, or about two English miles and a half in length, though part of the wire dragged upon wet grass, went over hedges, palisadoes, and grounds newly plowed up; they also found that when a wire of one thousand three hundred and nineteen feet, with its extremities brought near each other, was electrified, the electricity ceased at one end the moment it was taken off at the other.

The discoveries of the French occasioned this branch of electricity to be carried still further by the English philosophers: Dr. Watson, with the assistance of several gentlemen, made a great number of experiments, on what is called the electric circuit; the most extensive of which was made upon "Shooter's Hill, on the 14th of August, 1747, at a time when there had been but one shower of rain during the five preceding weeks. In this experiment, the wire, communicating with the iron rod which made the discharge, was six thousand seven hundred and thirty-two feet in length, and was supported all the way upon baked sticks, as was also the wire which communicated with the coating of the phial, which was three thousand eight hundred and sixty-eight feet long, and the observers were distant from each other two miles. The result of the explosion demonstrated, to the satisfaction of the gentlemen present, that the circuit performed by the electric matter was four miles: viz. two miles of wire, and two of dry ground, the space between the extremities of the wire, a distance which, without trial, as they justly observed, was too great to be credited. A gun was discharged at the instant of the explosion, and the observers had stop-watches in their hands to note the moment when they felt the shock; but, as far as they could distinguish, the time in which the electric matter performed that vast circuit might have been instantaneous."

Although the effects of electricity are viewed by the greater part of mankind as mere matter of curiosity, yet the man of science takes a nobler and more extended view of the subject; he is anxious to pry into the hidden laws of nature, and discover, if possible, what part this active agent is performing in the busy scene of creation; and in no instance has this laudable spirit of inquiry been more gratified than by the discovery to which I shall now solicit the attention of the reader.

That there is strong similarity between the effects of lightning and electricity was very early remarked by several electricians; but the method that proved, by experiment, that lightning and electricity were exactly the same agents, was proposed by the justly-celebrated Franklin. This philosopher first suggested the idea of drawing electricity from the clouds, by pointed conducting bodies; for he had often observed the great power pointed bodies had in collecting artificial electricity. The doctor, after having published his hypothesis concerning the sameness of electricity with the matter of lightning, "was waiting for the erection of a spire, in Philadelphia, to carry his views into execution, not imagining that a pointed rod, of a moderate height, would answer the purpose, when it occurred to him, that, by means of a common kite, he could have a readier and better access to the regions of thunder than by any spire whatever." The kite that Dr. Franklin intended for this purpose was prepared in the following manner: to the top of the kite a pointed wire was fixed, to collect the electricity from the clouds, which the doctor supposed would descend the hempen string by which the kite was to be suspended. To the lower end of the hempen string a small key was to be attached, and to this key a few feet of silk line, by which the kite was to be held when elevated; and as silk is a non-conductor of electricity, he conceived it would prevent the fluid that descended the hempen string from escaping to the earth. Dr. Franklin, having his kite thus prepared, took an opportunity, the first approaching thunder-storm, to put his plan into practice; but dreading the ridicule which too commonly attends unsuccessful attempts in scientific inquiries, he communicated his intended experiment to no one but his son, who assisted him in raising the kite.

"The kite being raised, a considerable time elapsed before there was any appearance of its being electrified. One very promising cloud passed over it without any effect; when, at length, just as he was beginning to despair of his contrivance, he observed some loose threads of the hempen string to stand erect, and avoid one another, just as if they had been suspended on a common conductor. Struck with this promising appearance, he immediately presented his knuckle to the key, and (let the reader judge of the exquisite pleasure he must have felt at that moment) the discovery was complete. He perceived a very evident electric spark; others succeeded, even before the string was wet, so as to put the matter past all dispute; and when the rain had wetted the string, he collected fire very copiously." By this means the doctor was enabled to charge phials at the key, and collect sufficient electric fire to perform many of those experiments which are usually exhibited by the electrical machine.

These results the doctor obtained in June, 1752, a month after the electricians in France had obtained the same, but before he had heard any thing of their experiments. "Besides this kite, Dr. Franklin had afterwards an insulated iron rod to draw the lightning into his house, in order to make experiments whenever there should be a considerable quantity of it in the atmosphere, and that he might not lose any opportunity of that nature, he connected two bells with this apparatus, which gave him notice, by their ringing, whenever his rod was electrified."

Mr. De Romas was the first that interwove a wire into the string of an electrical kite, and the kite this gentleman used was seven feet and a half high, and three feet wide: "By the help of this kite, on the 7th of June, 1753, about one in the afternoon, when it was raised 550 feet from the ground, and had taken 780 feet of string, making an angle of near 45 degrees with the horizon, he drew sparks from his conductor, three inches long, and a quarter of an inch thick, the snapping of which was heard about 200 paces." "The quantity of electric matter brought by this kite from the clouds, at another time, is really astonishing. On the 16th of August, 1756, the streams of fire issuing from it were observed to be an inch thick, and ten feet long. This amazing flash of lightning, the effect of which on buildings, or animal bodies, would perhaps have been equally destructive with any that are mentioned in history, was safely conducted by the cord of the kite to a non-electric body placed near it, and the report was equal to that of a pistol." But the great practical use of this discovery was to secure buildings from being damaged by lightning: "This great end Dr. Franklin accomplished by so easy a method, and by so cheap, and seemingly trifling apparatus, as fixing a pointed metalline rod higher than any part of the building, and communicating with the ground, or rather the nearest water. This wire the lightning was sure to seize upon, preferable to any other part of the building, whereby this dangerous power would be safely conducted to the earth and dissipated, without doing any harm to the building."

Having given a general account of the rise and progress of the most important discoveries in electricity, I shall now proceed to inquire into that part of this science which more particularly belongs to the practical electrician, and lay down some general rules by which those but little acquainted with the subject may perform the most interesting experiments in this extensive branch of natural philosophy.

Source Endnote

1 This gentleman was killed at Petersburgh by a flash of lightning, which he had incautiously invited into his room while making some experiments on atmospheric electricity.

Editorial Headnote

5. "Progress of the Mind, Canto III," from *The Temple of Nature or the Origin of Society* (London: Jones and Company, 1825), pp. 33–34.

The British natural philosopher, physician, abolitionist, tinkerer, and poet Erasmus Darwin is well known for his writings on natural history, evolution, and social issues.[1] He circulated among the most erudite circles in the natural philosophical community, including the Royal Society and the less well-known Lunar Society. In one of his most notorious and final works, *The Temple of Nature*, first published in 1803, Darwin colorfully recasts the development of mankind through the eyes of Urania, the Muse of Science. Most notably, he argues that life emerged from the primeval sea, which many connect to the work of his grandson Charles Darwin in *The Origin of Species*. He even speculates about the construction of automata birds. In the selection included here, from the third Canto of the poem, the muse explains the growth of intellectual and moral faculties. In these opening lines, Darwin invokes "illuminated," "electric streams," "resinous and vitreous fire," and "Galvanic chain-work." The words not only elicit figurative meanings, including something that is intense, brilliant, or thrilling but also display Darwin's familiarity with the notable experiments of people like Luigi Galvani (1737–1798) that so fascinated early-nineteenth-century natural philosophers. Furthermore, it highlights how electricity as object of study spread beyond scientific halls and into literature. Electricity was both a burgeoning field of science and mechanism through which many understood life to originate.

Editorial Endnote

1 Paul A. Elliot, *Erasmus Darwin's Gardens: Medicine, Agriculture and the Enlightenment Sciences* (Boydell, 2021), 6.

5

PROGRESS OF THE MIND, CANTO III

Erasmus Darwin

Source: Erasmus Darwin, "Progress of the Mind, Canto III," from *The Temple of Nature or the Origin of Society* (London: Jones and Company, 1825), pp. 33–34.

I. Now rose, adorn'd with Beauty's brightest hues,
The graceful Hierophant, and winged Muse;
Onward they step around the stately piles,
O'er porcelain floors, through laqueated aisles,
Eye Nature's lofty and her lowly seats,
Her gorgeous palaces, and green retreats,
Pervade her labyrinths with unerring tread,
And leave for future guests a guiding thread.

First with fond gaze blue fields of air they sweep,
Or pierce the briny chambers of the deep;
Earth's burning line, and icy poles explore,
Her fertile surface, and her caves of ore;
Or mark how Oxygen with Azote-Gas
Plays round the globe in one aerial mass,
Or fused with Hydrogen in ceaseless flow
Forms the wide waves, which foam and roll below.

Next with illumined hands through prisms bright
Pleased they untwist the sevenfold threads of light;
Or, bent in pencils by the lens, convey
To one bright point the silver hairs of Day.
Then mark how two electric streams conspire
To form the resinous and vitreous fire;
Beneath the waves the fierce Gymnotus arm,
And give Torpedo his benumbing charm;
Or, through Galvanic chain-work as they pass,
Convert the kindling water into gas.

DOI: 10.4324/9781003295556-7

How at the poles opposing Ethers dwell,
Attract the quivering needle, or repel.
How Gravitation by immortal laws
Surrounding matter to a centre draws;
How Heat, pervading oceans, airs, and lands,
With force uncheck'd the mighty mass expands;
And last how born in elemental strife
Beam'd the first spark, and lighten'd into Life.

Now in sweet tones the inquiring Muse express'd
Her ardent wish; and thus the Fair address'd.
"Priestess of Nature! whose exploring sight
Pierces the realms of Chaos and of Night;
Of space unmeasured marks the first and last,
Of endless time the present, future, past;
Immortal Guide! O, now with accents kind
Give to my ear the progress of the Mind.
How loves, and tastes, and sympathies commence
From evanescent notices of sense?
How from the yielding touch and rolling eyes
The piles immense of human science rise? –
With mind gigantic steps the puny Elf,
And weighs and measures all things but himself!"

The indulgent Beauty hears the grateful Muse,
Smiles on her pupil, and her task renews.
Attentive Nymphs in sparkling squadrons throng,
And choral Virgins listen to the song;
Pleased Fawns and Naiads crowd in silent rings,
And hovering Cupids stretch their purple wings.

II. "First the new actions of the excited sense,
Urged by appulses from without commence;
With these exertions pain or pleasure springs,
And forms perceptions of external things.
Thus, when illumined by the solar beams,
Yon waving woods, green lawns, and sparkling streams,
In one bright point by rays converging lie
Plann'd on the moving tablet of the eye;
The mind obeys the silver goads of light,
And Irritation moves the nerves of sight.

"These acts repeated rise from joys or pains,
And swell Imagination's flowing trains;

So in dread dreams amid the silent night
Grim spectre-forms the shuddering sense affright;
Or Beauty's idol-image, as it moves,
Charms the closed eye with graces, smiles, and loves;
Each passing form the pausing heart delights,
And young Sensation every nerve excites.

"Oft from sensation quick Volition springs,
When pleasure thrills us, or when anguish stings;
Hence Recollection calls with voice sublime
Immersed ideas from the wrecks of Time,
With potent charm in lucid trains displays
Eventful stories of forgotten days.
Hence Reason's efforts good with ill contrast,
Compare the present, future, and the past;

Editorial Headnote

6. "Electricity," *Saturday Magazine* 6: 169, February 21, 1835, pp. 68–69.

Between 1832 and 1844, the Committee of General Literature and Education, a Society for Promoting Christian Knowledge subsidiary, published the *Saturday Magazine* to compete with the famous *Penny Magazine* and *Chamber's Edinburgh Journal*.[1] Each of these publications sold for a penny, contained numerous illustrations, and hoped to appeal to the "intellectual working class."[2] In their pages, editors spoke of major scientific discoveries of the day, the latest political debates, and notes on the happenings around the British Empire. They succeeded to varying degrees as part of a number of political reforms targeting education, but their audience came to prefer the publication of stories and literature. In the section cited here, the magazine published a short piece on "Electricity" which captures the excitement and potential people associated with electrical science and demonstration during this period. Additionally, the article nods to one of Michael Faraday's recent speculations about the "electric equilibrium of the earth" as being responsible for the Aurora Borealis.[3] The rest of the article ranges in topics from meteorology, geology, physics, and biology but all of them remained connected to the wonders of electricity that the editors and their audience held in the early decades of the nineteenth century. They imagined a substance of limitless potential that tied together the branches of natural philosophy.

Editorial Endnotes

1 Johnathan R. Topham, "The Mirror of Literature, Amusement and Instruction and cheap miscellanies in early nineteenth century Britain," in Cantor, Dawson, Gooday, Noakes, Shuttleworth, and Topham, *Science in the Nineteenth-Century* Periodical (Cambridge: Cambridge University Press, 2004), 37–40.

2 Richard Salmon, "The Unaccredited Hero": *Alton Locked*, Thomas Carlyle, and the Formation of the Working-Class Intellectual," in Aruna Krishnamurthy, *The Working-Class Intellectual in Eighteenth and Nineteenth Century Britain* (Burlington, VT: Ashgate, 2009), 167–169.

3 Michael Faraday, "The Bakerian Lecture: Experimental Researches in Electricity," *Philosophical Transactions of the Royal Society* 22 (1832): 163–194.

6

ELECTRICITY

Source: "Electricity," *Saturday Magazine* 6: 169, February 21, 1835, pp. 68–69.

ELECTRICITY

Electricity is one of those agents pervading the earth and all substances, without giving any visible sign of its existence, when torpid; yet, when active, often producing violent and destructive effects.

Electricity may be called into action, by mechanical, chemical, and magnetic powers, and by heat; but we are ignorant how it is roused, or of the manner of its existence in bodies. It is *supposed* to be a highly elastic fluid, capable of moving through matter; and as experience shows that bodies in one electric state attract, and in another repel, each other, the system of positive and negative electricity is adopted. As each electricity has its peculiar properties, the science may be divided into the following branches.

Substances in which the two electricities are combined, neither attract nor repel, except when the electric equilibrium is destroyed by friction; then the positive electricity is impelled in one direction, and the negative in another. Those of the same kind repel, and those of different kinds attract, each other. When the attractive and repulsive powers are not opposed at equal distances, they join suddenly with great rapidity, and produce the electric flash, explosion, and shock; then the natural state is restored, and electricity remains quiet till called forth by a new cause.

When two plates of glass, the one polished, and the other rough, are rubbed against each other, the polished surface acquires positive, and the rough negative electricity. Equal lengths of black and white ribbon, drawn together between the finger and thumb, so as to rub their surfaces together, the black ribbon has acquired negative, and the white, positive electricity but if the whole length of the black ribbon be drawn across the white, the electricities will be changed; the black positive, the white negative.

Electricity passes through all substances, but with more facility through metals, water, the human body, & c., which are called conductors, than through air, glass, and silk, which are called non-conductors. When bodies are not surrounded with non-conductors, the electricity escapes quickly into the earth; a body charged with electricity, and surrounded by non-conductors, so that the

DOI: 10.4324/9781003295556-8

electric fluid shall not escape, produces electricity of an opposite kind, in all bodies near it, and a state similar to its own in the distant parts; so that the two opposite electricities will attract each other.

Electricity is confined to the surface of bodies, as the exterior may be electric, and the interior in a state of neutrality. When electric matter is restrained by the non-conducting power of the air, which opposes its escape, and if the pressure of the electric matter is less than the force of the air, it is in a state of quiet; but when the air is extended by moisture, and it exceeds that force in any one point, the electricity escapes. Electricity is more easily retained by a sphere, but escapes from a point, and a pointed object receives it with most facility.

The heat produced by an electric shock is intense, but it is only accompanied by light, when the fluid is obstructed in its passage.

Electrical light is the same as the light of the sun, it seems to arise from the air being condensed during the rapid motion of the electric fluid; the production and condensation of vapour is a great source of atmospheric electricity. The atmosphere, when clear, is positively electric, and stronger in winter than in summer, and during the day than in the night.

Atmospheric electricity arises from the flowing of the electric fluid during the evaporation that arises from the earth. Clouds owe their existence and form probably to it, as they consist of hollow tubes, coated with electricity, and as it is either positive or negative the tubes repel each other, which prevents their uniting and falling down in rain. When two clouds charged with opposite kinds of electricity, approach within a certain distance, the thickness of the coating of electricity will increase on the two sides of the clouds that are nearest to one another, and when the accumulation becomes so great as to overcome the pressure of the atmosphere, a discharge takes place which occasions a flash of lightning. The pure air, at all times negatively electric, becomes more so on the approach of rain, snow, wind, or hail; but it afterwards varies on opposite sides, and the transitions are very rapid on the approach of a thunder-storm.

Editorial Headnote

7. *The Silent Revolution, Or, The Future Effects of Steam and Electricity Upon the Condition of Mankind* (London: William and Frederich G. Cash, 1852), pp. 1–13.

Garvey's text is a forward-looking meditation on how technological advancements, particularly steam and electricity, would shape the future of humanity. Although not a well-known author, the way he envisioned a world profoundly altered by these forces, which he believes would democratize power and improve living conditions, represented broader Victorian ideas about the positive impact of electricity.[1] As industrial processes transformed Britain's economy and society, Garvey's work is emblematic of the broader Victorian fascination with technological progress and its utopian possibilities.[2] He argued not just about the practical uses of steam and electricity but also about their potential to revolutionize social structures. Garvey saw these technologies as agents of social change that would bridge economic divides, enhance communication, and increase human productivity. He imagined a future in which the disparities between rich and poor are mitigated by the efficient distribution of power and resources, and where the manual labor of the working class is alleviated by mechanical processes. Although many historians have shown that electric systems often replicated restrictive political and economic structures, idealism like Garvey's persisted.

In the context of Victorian imaginings of electric power, Garvey's work anticipated the later enthusiasm for electricity as a symbol of modernity, control, and boundless energy. *The Silent Revolution* contributes to the growing narrative of electricity as a liberating force, paralleling the later works of figures like Silvanus P. Thompson and William Thomson.[3] It fits within the broader framework of Victorian optimism about scientific progress, which envisioned electricity not only as a technological tool but also as a means to transcend the limitations of the human condition. Garvey's vision of a future shaped by steam and electricity reflects the nineteenth-century belief that these technologies could catalyze social and moral improvement, thus situating *The Silent Revolution* within the history of utopian thought and early electrical imagination.

Editorial Endnotes

1 Stella Pratt-Smith, *Transformations of Electricity in Nineteenth-Century Literature and Science* (Routledge, 2016), 147.
2 Iwan Morus, "In the Ether: Electricity and the Victorian Future," in Utopianism and the Sciences, Kemperink, Mary; Vermeer, Leonieke eds. (Peeters Publishers, 2010) 17–32.
3 Crosbie Smith and Norton Wise, Energy and empire: A biographical study of Lord Kelvin (Cambridge, 1989), 150; Stanley, M. (2021). No slaves to words: S. P. Thompson's theory of history. *Centaurus*, 63(3), 489–498.

7

THE SILENT REVOLUTION, OR, THE FUTURE EFFECTS OF STEAM AND ELECTRICITY UPON THE CONDITION OF MANKIND

Michael Angelo Garvey

Source: Michael Angelo Garvey, *The Silent Revolution, Or, The Future Effects of Steam and Electricity Upon the Condition of Mankind* (London: William and Frederich G. Cash, 1852), pp. 1–13.

THE SILENT REVOLUTION,

&C.

CHAPTER I

No one can contemplate the unexampled progress of science within the present century, without feeling that a new epoch has commenced in the history of our race. The divine powers of the human mind are extending their grasp and rising to a state of higher activity. Fields of knowledge undreamt of in the earlier ages of the world are successfully cultivated. The farthest regions of space are explored, and the secrets of their starry depths unfolded to men. The hidden forces of nature; the laws by which her phenomena are governed in their endless variety and succession; the economy of being; the structure and properties of matter; the relations of things and of ideas; the very mind itself: – all are undergoing a rigorous process of scientific investigation, from which result discoveries, that would be deemed miraculous, did not their number and frequency almost exhaust our faculty of wonder.

The great characteristic of this general intellectual movement is its practical tendency. The public in our day expects from its thinkers and experimentalists, not clever paradoxes, moving sentimentalisms, nor ingenious puzzles, but plain demonstrations of the best method of grappling with obstinate realities; discoveries of the means by which the powers of nature may be brought to subdue one another to the service of man in his ceaseless struggle with her material elements. Learning and skill are esteemed in proportion as they conduce to the well-being of society, and their value is measured by the extent of their application to the

DOI: 10.4324/9781003295556-9

practical affairs of life. The genius of the age has responded to this expectancy. Philosophers have ceased to speculate in retirement upon unsubstantial and useless theories. Science is no longer a lifeless abstraction floating above the heads of the multitude. It has descended to earth. It mingles with men. It penetrates our mines. It enters our workshops. It speeds along with the iron courser of the rail. It tramples on the billows. It defies the tempest. It gives to man the sunbeam for a pencil, and the lightning for a messenger. It lends to his feeble arm an irresistible might before which mountains crumble into dust; the barriers of kingdoms are removed, estuaries and straits are spanned with substantial road-ways, and the unstable waters, no less than the firm land, are subjected to his dominion. The talismans of Arabian fable never endowed their possessors with such power as that which science has bestowed upon mankind. It has placed at our command agencies, whose indefatigable energy, and adaptation to all human purposes, cast into shade even the genius of the ring or the lamp. What is there in all the recorded dreams of fancy more wonderful than the force of steam in its various applications? How stupendous in its power, and yet how manageable! A child may direct it. It would crush an army. A single touch puts in motion the majestic ship laden with the wealth of a nation. The rushing train, freighted with a thousand lives, or the complicated machinery that groans and whirls throughout the town-like factory, and a touch equally slight arrests the mighty power. In one place we see it wielding without effort a Nasmyth hammer of many tons weight, and forging the ponderous anchors destined to hold the floating castles of England against the fury of the hurricane. In another, polishing with an accuracy, beyond human skill, the delicate hairspring that is to trace and record the noiseless progress of time. Here it is employed upon massive blocks of iron, which it rolls out, cuts up, and moulds as the potter does the clay; and here it is spinning threads so fine that they almost elude the sight, and weaving them into airy textures that look like wreaths of morning mist, Unaffected by place, time or climate, incapable of fatigue, untouched by passions or infirmities, there stands the universal servant of man, ready to relieve him from all drudgery, and to supplement his limited ability in carrying out the intentions of his will. It matters not how difficult or various the services required, nor where they are to be performed. In the depths of the earth, or on the mountain top; in the open field, or in the crowded city; in the frozen north, or the burning tropics; whether they require the most gigantic strength, or the nicest care, this wondrous agent is suited to them all. It enables man, who is slow and weak compared with other terrestrial creatures, to pass from place to place with the speed of the eagle, carrying burdens in his flight that would crush the strong elephant to dust. By its means the force and dexterity of a million fingers are subjected to the control of one mind, and imbued with its intelligence. Under the transforming touch of this marvel-working power, the rudest substances assume forms of beauty and utility. The dark shapeless ore divides itself into multitudinous forms subservient to human purposes; it sparkles in the gorgeous service of the palace; it ministers to the humble comforts of the cottage; it gives instruments to the philosopher, and tools to the artisan; it furnishes man with means and appliances for executing his largest designs, and securing his conquest of the whole earth. The frail vesture of the cotton seed, that once rotted unnoticed where it fell, becomes clothing for nations, and when it has answered all possible purposes, and been reduced to the state of filthy worn-out rags which the very beggar on the high-way casts off with disgust, even these are taken up, and, by a magical process, transmuted into fair pages which are impressed with imperishable thoughts, multiplied with a rapidity like that of thought itself, and distributed throughout the world. Locomotion by steam is the complement of all its other applications; it has broken down the boundaries which time and space had set to human enterprise; it has emancipated the genius of our race from local restrictions, made the wonders wrought in any one place the common property of mankind, and combined the material interests of all nations. These are the miracles of steam, of the latent power in that impalpable vapour which has played before mankind since the creation, but which it was reserved for our age to control and subjugate. Science has taught us the spell by which the strong spirit of the mist is conjured from his dark abode in the mine, and compelled to execute our behests before he escape into the regions of air. The imponderable elements of nature are more powerful than her most obdurate substances. Xerxes tried in vain, with the strongest materials and the labour of myriads, to bridge the narrow Bosphorus; a single winter destroyed the vast structure: with us the most distant portions of the world are united by the invisible force that forms the summer cloud. But even the subtle influence of caloric, which thus animates the inert waters into titanic might, fails to excite our astonishment to so high a degree as electricity. If the one unites lands and transports material objects, the other seems destined to become the link of minds, the channel of intelligence and thought between all habitable parts of the globe. As yet, we can hardly venture to conjecture the form this mysterious power will assume, or the part it will perform in the future history of the world. Some general ideas are all we can fix upon as the basis of our speculations; but these are amongst the most astounding facts within the compass of human knowledge. They resemble the operations of a spiritual, rather than of a material agency, and the recondite manner in which they are effected entirely eludes our investigations. The electric telegraph is now a familiar object; we look upon it with a careless eye, we pass onward, and forget if we ever considered its miraculous properties. Yet, here we see realized more than the beautiful fiction of Strada; not between two minds alone does it establish a responsive sympathy, it creates a communion of thought between cities and kingdoms; it enables a man to utter his feelings, at the very instant they arise in his mind, to ears that listen for them at a distance of a thousand miles. What is this surprising and intangible power? What will it accomplish in its maturity, when it so far transcends all calculation in its infancy? One might think for ever, with increasing wonder, on the one supernatural fact, that without any perturbation of its parts perceptible to the finest organs, or appreciable by the most exquisite instruments, communications are transmitted along the wire, however great its length, with such inconceivable rapidity, that the moment of transit cannot be measured by the most nicely adjusted chronometers.

There is no sensible lapse of time between the giving of the signal at one terminus, and its reception at the other, though the distance between them should be a thousand, ten thousand, or twenty thousand miles. Nay, if a wire were stretched round the whole earth, and the ends made to terminate in two dials placed immediately beneath the eye, the very instant the index that supplied the influence was moved, it would be replied to by that which received it after traversing twenty-five thousand miles of space; the movements would be to all intents and purposes simultaneous, and this result would be unaffected by the nature of the elements through which the conducting medium might pass. It would be still the same thought rivalling speed, whether the conductor were raised in the air or sunk in the earth, whether it passed over the loftiest mountain peaks, or lay at the bottom of the ocean with the world of waters rolling above. Whilst we look upon the electric wires as they stretch along by the great Unes of railway, they appear perfectly quiescent. The weary little bird rests upon them in his flight, and clasps them in his tiny claws; yet along that motionless thread, and through that feeble grasp, there may be passing tidings of life or death, of ruin or prosperity – intelligence of the fall of kings and thrones, of battles lost and won, of events that change the destinies of millions, and plunge whole nations into mourning or intoxicate them with joy. A thousand fathoms beneath the keel of the war-ship, undisturbed by the tumult of the elements in which she reels and struggles; in the dark and silent abysses of ocean, where uncouth monsters make their retreat, and human vision has never penetrated; amidst objects strange to the eye of man, and scenes that have been secret since the beginning of time, there lies the wondrous ligature which connects the minds of nations, conveying manifold contributions to the sum of human wisdom and experience, from which man shall yet learn to still his mimic thunders, and to aspire after higher glories than those won by mutual slaughter, on the bloody waters, or the groaning field. Such are the new agencies whose influence upon the future condition of mankind we have to consider. But there is another power that incalculably transcends them, an agency to which steam and electricity and all other physical forces are merely means and instruments. It is that of the mind itself, the prime mover and director of them all, which defines their purposes, and controls their operations – that restless, indefatigable, and inscrutable principle to which even the human frame is but an apparatus of exquisitely adjusted organs – that intellectual power, which, as it feels, and thinks, and wills, impels man to action, designs his objects, directs his movements, and registers their results; which pushes its inquiries into every department of creation, and with the same insatiable thirst for knowledge scans the mightiest operations of nature, and scrutinizes her minutest processes; which surveys the past, anticipates the future, and ever and anon turns in upon itself to observe its own movements, and to conduct a marvellous analysis of which it is itself at once the subject, the instrument, and the performer – the immortal element in man's nature which tinges with the glow of its own etherial essence his feeble and delicate frame, investing him with sovereign dignity, and giving him power over all terrestrial things and beings.

To the study of this principle we must inevitably recur, if we would either trace with accuracy the progress of our race through the past, or indicate the probable course of its future destinies. It is the fountain of all human activity, and the final issues of all man's labour and discovery depend upon its character and tendencies. We propose to estimate these tendencies and to ascertain the conditions under which their development is either retarded or promoted, in order that we may be able to form some rational conjecture of the future of our species. This would be a difficult and uncertain task, if attempted from an abstract consideration only of the human faculties. There is, however, a sure and easy method of accomplishing that object. The path which man has already trodden from barbarism to civilisation, lies before us in authentic history, strewed with the memorials of his passions and his might. There we can calmly investigate his progress, note the difficulties and obstructions it has had to overcome, and deduce with certainty the principles that have given it its forward impulse, as well as the conditions under which that impulse has acted with greatest power and constancy. From the study of what man has accomplished, we shall rise with a fuller conception of what he is likely to accomplish, for the past is a dim mirror of the future, displaying in shadowy twilight the vague outline of that wondrous phantasmagoria in which he is the chief figure, and which is gradually rising into distinctness, as the great secular morning brightens into day. The principles of human action which have issued in the events of foregone history, are in their nature immutable, and will create the history of the times to come. The ages that have fled resemble those that are approaching, as the wilderness or partially reclaimed field resembles the highly cultivated garden. Their products are different, but the powers of the soil are the same, and all that art has done, has been to give those powers liberty of development. The same vital principles of life and growth that once clothed the desert with briars and thorns, now sustain the fragrant rose and the tree laden with golden fruit. In like manner the necessities, passions, and affections of our nature are constant in their operation, and from them spring all our actions; they are the very pabulum and support of human purposes and efforts. What we call social laws, or the principles upon which societies of men arise and are held together, are but the resultants of those innumerable minute impulses which are hourly and momentarily springing up in the breasts of individuals. To arrive at just conclusions, then, concerning the direction in which our race is moving, it will be necessary to consider this compound force by which it is impelled; this will imbue us with true notions of the subjective energy which has enabled mankind to attain its present degree of elevation; whilst, by tracing its previous course, the mind will acquire a corresponding momentum which will project it forward in the right line of conjecture as to the point at which that course will culminate, though its incidents remain hidden from us amongst the secrets of time.

Editorial Headnote

8. "Science," *The Westminster and Foreign Quarterly Review*, January 1856, pp. 254–271.

Beginning in 1824, the British quarterly publication *Westminster Review* functioned as one of the organs of the Philosophical Radicals, comprised of philosophers such as Jeremy Bentham (1748–1832), James Mill (1773–1836), Francis Place (1771–1854), and John Stuart Mill (1806–1873). By the time of this essay, the publication had continued writing and pushing their radical rhetoric and began to attach many of their ideas of progress to major changes in science and the mechanical arts, especially after 1853 with the addition of John Tyndall (1820–1893) and Thomas Huxley (1825–1895) as editors of the "Science Section." Included here is the science section from January 1856. In this issue, the editors focused on Michael Faraday's (1791–1867) third volume of *Experimental Researches in Electricity* and review a number of other texts.[1] Their high opinion of the work and Faraday's general contributions to science are evident as they discuss his light, magnetism, and experimentation, among many other things. What is perhaps most notable is the position in which they place Faraday and the science of electricity. Not only are they a matter of extreme interest to the scientific minds of the day, but they also provide direction. For the editors, and undoubtedly for many of the readers, electricity had the most potential to contribute to the improvement of society. Explanations of electric phenomena evaded the precise description present in other sciences and therefore experiments that offered further insight or mechanism were prized. Like all Nature in the minds of Victorians, electricity was to be tamed and made subservient. Electricity, even outside of explicitly scientific publications, captured the hopeful imagination of the intellectual and middle classes of Victorian Britain. In the opening pages of this essay, the editors make it clear that they saw electricity as a vehicle for the improvement of science and society.

Editorial Endnote

1 Michael Faraday, *Experimental Researches in Electricity* (London: Taylor and Francis, 1855).

8

SCIENCE

Source: "Science," *The Westminster and Foreign Quarterly Review*, January 1856, pp. 254–271.

FARADAY'S experimental researches are unquestionably among the most important that the century has produced; and of these researches those described in the volume just published[1] are amongst the most interesting, inasmuch as they have revealed the existence of an entirely new branch of physics, and given a direction to the ablest scientific thought of our day.

It is difficult to determine in what respect this third volume will be most valuable to us, for, in addition to the important results of his investigations, Faraday's method of investigating is exceedingly instructive. As an experimentalist he is, perhaps, unequalled. His descriptions of experiments, too, are always lucid and admirable, and this volume is full of them. Their ingenuity and simplicity, and the conscientious accuracy with which they are interpreted, are as charming to the general reader as they are instructive to the student. Viewed merely as lessons in experimentation, therefore, Faraday's researches are invaluable, if not most valuable.

The third volume commences with the celebrated memoir "On the Magnetization of Light, and the Illumination of Magnetic Lines of Force," the title of which requires a little explanation in order to render it generally intelligible. It is well known that a ray of light falling upon a plane mirror at an angle of incidence equal to about 54° 35′ is polarized by reflection, and that the polarized ray, when received upon a second mirror at the same angle of incidence, cannot be reflected with equal facility on all sides. If the second mirror be turned round the ray as an axis, so as to preserve the same inclination towards it, and a screen be held so as to receive the reflected ray, the corresponding image is most distinct when the mirror occupies either of two determinate positions 180° distant from each other, it diminishes when the mirror is turned out of these positions, and is finally extinguished in two positions at right angles to the first. This fact is expressed by saying that the ray reflected from the second mirror is most intense when the plane of incidence and reflection coincides with the plane of polarization, and is extinguished when the former plane is at right angles to the latter – the plane of polarization being that which coincides with the plane of incidence on the first mirror.

If a prism of heavy glass, or silicated borate of lead, be placed between the two mirrors, so that the polarized ray shall pass along its axis, the same position of the second mirror as before will be found to extinguish the ray, thereby proving that the position of its plane of polarization has not been altered in its passage through the heavy glass. The second mirror being placed so as to extinguish the ray, let a powerful horse-shoemagnet be brought close to the bar of heavy glass so that the latter occupies the space between its poles, or, more strictly, so that the line joining the poles coincides as nearly as possible with the direction of the ray, without obstructing its passage. The image will be immediately restored; and, in order to extinguish it again, the second mirror must be turned on its axis in a certain direction. This proves that by the influence of magnetism the plane of polarization has been made to rotate in that direction. If the magnet be withdrawn, the plane of polarization will return to its original position; if its poles are made to change places, an equal rotation in an opposite direction will be imparted to the plane of polarization. Everything else being the same, the magnitude of rotation will increase with the strength of the magnet, and with the length of the bar of heavy glass traversed by the ray, and will decrease as the angle made by the line joining the magnetic poles and the ray increases; when this angle is equal to 90°, the heavy glass has no action on the ray.

Such is the simplest form of the experiment made by Faraday, and which he called the magnetization of light. Besides heavy glass, many other bodies manifested the same action when exposed to magnetic influence. The results of all the experiments, which were published in 1845, are of great importance in three ways. First, they establish a relation between magnetism and light. Secondly, they may enable us to discover the nature of that peculiar power which some bodies naturally possess, of causing the plane of polarization to rotate. Thirdly, they establish the general fact that bodies, formerly thought to be unaffected by magnetism, are not so in reality.

In the two first directions, little progress has been made since Faraday's discovery; in the last, great progress has been made. The above experiments suggested an examination of the action of magnets upon *diamagnetic* bodies generally; in other words, upon bodies which were not acted upon by magnets in the manner of iron or loadstone. Hence arose the new branch of science known as diamagnetism. The same prism of heavy glass which had first manifested an action upon light, and which was about half an inch square and two inches long, was suspended horizontally, by means of a silken thread, between the poles of a powerful electro-magnet. When the magnet was excited, the prism turned and set its length *equatorial* or perpendicular to the line joining the poles, and when forcibly withdrawn from this position it always returned to it. When suspended close to a single pole, it was always repelled *en masse,* and set its length perpendicular to the resultant of magnetic force, or to the line joining its centre and the pole. It is scarcely necessary to say, that a similarly shaped magnetic body (*e.g.* soft iron) would, in the first experiment, have set its longest dimension *axial,* or from pole to pole, and in the second experiment, it would have been attracted *en masse.* Every other body, when reduced to a suitable form, and placed between the poles of the magnet, was found to behave either like heavy glass or like soft iron, and thus a method was obtained for dividing all bodies into two classes – the magnetic and diamagnetic, of which classes the latter was by far the most numerous.

Amongst solids, glass, phosphorus, sulphur, resin, sealing-wax, caoutchouc, starch, wood, ivory, all kinds of flesh, leather, bread, and all the common metals, with the exception of iron, nickel, cobalt, and manganese, were found to be diamagnetic. If a man could be suspended between the magnetic poles, he would be found to point equatorially, for all the constituents of his body, his *blood* included, are diamagnetic. Amongst diamagnetic metals, bismuth manifests the most energetic action, so that in diamagnetic experiments it plays an important part.

A great many liquids were examined by enclosing them in tubes of thin glass, which, although itself diamagnetic, obscures but little the action of any liquids the tubes may contain. In this manner water, alcohol, ether, nitric and other acids, solutions of various alkaline and earthy salts, olive oil, oil of turpentine, and blood, were all proved to be dia-magnetic. Solutions of the salts of iron, on the contrary, were magnetic.

Experiments on gases, made by enclosing them in glass tubes hermetically sealed, did not at first succeed; the diamagnetism of the tubes masking any action the gases might possess. More delicate experiments, however, made by Faraday in 1850, in which the effect of the glass was eliminated, established the magnetic and diamagnetic properties of several gases. Oxygen is strongly magnetic, and the more so the denser it is. Atmospheric air, therefore, is magnetic, for its other constituent, nitrogen, appeared to be neutral.

Finally, many extremely instructive results were obtained by observing the action of bodies when immersed in different media, liquid and gaseous, occupying the magnetic field. Substances pointed axially between two poles, and were attracted by one pole, when they were immersed in media less magnetic or more diamagnetic than themselves, whilst they pointed equatorially, and were repelled, when immersed in media more magnetic or less diamagnetie than themselves. When we reflect that the experiments which led to the classification of bodies were all made in air, which is known to be a magnetic medium, the idea at once suggests itself, may not all these actions which were then observed have been of this *differential* kind; may not all bodies be magnetic, and may not those which appear to be diamagnetic be merely *less* magnetic than air, in other words, may not *attraction* be the only effect of the magnetic force, as we know it to be, of the force of gravity? The analogy between the two cases will be evident. A piece of cork in a vacuum, or in air, falls towards the earth's surface, whereas, when immersed in water, it is apparently repelled from the same. The experiment in vacuo suggests the explanation of the actions in air and water. The cork is always attracted by the earth, and the air it displaces is less attracted, whereas the water it displaces is more attracted than itself. May not apparent magnetic attractions and repulsions be similarly explained? Faraday thinks not, or rather he thought not at the time to which we refer, for in a vacuum, bodies are repelled as well as attracted by a magnet, and, rather than attribute magnetic properties to mere space, he then preferred the idea of the existence of repulsion as well as attraction, which, it is needless to say, is in perfect accordance with the differential actions just described. At present, however, Faraday is inclined to believe in the existence of a magnetic medium in space.

With regard to the explanation of the movements of diamagnetic bodies, Faraday suggested that magnets, by induction, might cause in them an opposite state to what they produced in magnetic bodies. Thus he supposed that the pole of a magnet which induces an *unlike* polarity in the nearest extremity of a magnetic particle, might induce a *like* polarity in the nearest points of a diamagnetic particle. This being admitted, the fundamental laws of magnetic action would explain all the motions of diamagnetic, as well as those of magnetic bodies.

An interval of nearly three years elapsed before Faraday published his next series of researches. In the meantime the whole scientific world had read of, and corroborated his experiments. Amongst those who contributed principally to the advancement of the science during this interval, Plücker, Reich, Weber, Poggendorff, and Edmond Becquerel, may be mentioned. The researches of most of them were principally directed towards the establishment of Faraday's suggestion, that "diamagnetic bodies possess a polarity the same in kind, but opposite in direction to that possessed by magnetic ones."

Weber asserted that, by means of a bar of bismuth which was exposed to magnetic action, he had produced induced currents of electricity exactly opposite in direction to those which would have been induced had iron been used instead of bismuth. The currents with bismuth, however, are so weak, and so many other sources of error are involved in this experiment, that it gave but little satisfaction at the time. Upon trial in 1850, Faraday himself could not obtain Weber's results, and, on that account, the former felt himself compelled to withdraw his first opinion on diamagnetic polarity. We may here state, in order not to have to return to the subject, that Weber shortly afterwards repeated his experiments in a far more accurate and unexceptionable form, and corroborated his former results; indeed, he succeeded in an approximate quantitative determination of the magneto-electric power of bismuth. He also showed that an electro-diamagnet could be formed by sending currents of electricity around a bar of bismuth, just as an electro-magnet is made by sending currents around a bar of iron, and that, all other things being the same, the polarity of the bismuth is exactly opposite to that of the iron. More recently, Tyndall, by numerous experiments, has more than corroborated Weber's results, for he has shown that, "whatever title to polarity the deportment of a bar of soft iron surrounded by an electric current, and acted on by other magnets, gives to this substance, a bar of bismuth possesses precisely the same title;"[2] so that if the discovery of diamagnetism could have preceded that of magnetism, diamagnetic polarity would have been the rule, and magnetic polarity the exception.

In 1848, at the close of the interval of three years above-mentioned, Faraday published some anomalous results which he had obtained with *crystalline* magnetic and diamagnetic bodies. For instance, he found that certain specimens of crystalline bismuth, without ceasing to be repelled *en masse* by either magnetic pole, set their longest dimensions *axial* instead of equatorial, and that crystals of sulphate of iron, without ceasing to be attracted *en masse,* sometimes placed themselves with their longest dimensions *equatorial* instead of axial. On further examination he found that in most crystals, magnetic as well as diamagnetic, there was a certain line or direction, connected in some manner with crystalline structure, which always set itself axial, although in doing so it might be opposed by the magnetic or diamagnetic tendency of the body consequent upon its form. This line in the crystal was called by Faraday its *magne-crystallic axis,* and to account for the axial position of the line, he supposed the existence of a new force – the *magne-crystallic force* – which was *directive* merely, and entirely independent of attraction and repulsion. In 1850, some experiments on gases led Faraday to the hypothesis of a magnetic conduction, somewhat similar to electric conduction, and he then thought that the above magne-crystallic actions might be explained by assuming that crystalline bodies conduct onwards, or permit the exertion of the magnetic force with more facility in the direction of the magne-crystallic axis than in any other.

In 1847, a year before the publication of Faraday's experiments on magne-crystallic action, Plücker, of Bonn, was also experimenting upon the deportment of crystals in the magnetic field, and endeavouring to establish another relation between magnetism and light. The results of his experiments led Plücker to the enunciation of the following two laws: –

> "When any crystal whatever, with one optical axis, is brought between the poles of a magnet, the axis is repelled by each of the poles, and if the crystal possess two axes, each of these is repelled with the same force by the two poles."
>
> "The force which causes this repulsion is independent of the magnetism or diamagnetism of the mass of the crystal. It decreases with the distance more slowly than the magnetic influence exerted by the poles."

Some time afterwards, in a letter to Faraday, these laws were greatly modified by their author. It was then affirmed, that "there will be either repulsion or attraction of the optic axes by the poles of the magnet, according to the crystalline structure of the crystal. If the crystal is a negative one, there will be repulsion; if it is a positive one, there will be attraction."

A similarity between Plücker's and Faraday's results will be at once perceived. The latter repeated the former's experiments, and concurred in his views. The effects, first observed by Plücker, were attributed by Faraday to the action of another new force, the *optic-axis* force, somewhat different from, though closely related to his own magne-crystallic force.

The subject remained in, this complicated state until Tyndall, in 1850, commenced a thorough investigation of the whole phenomena. Numerous experiments, remarkable for their ingenuity and simplicity, led to the enunciation of the following simple principle, by means of which all Faraday's and Plücker's results appeared capable of explanation.

> "If the arrangement of the component particles of any body be such as to present different degrees of proximity in different directions, then the line of closest proximity, other circumstances being equal, will be that chosen by the respective forces for the exhibition of their greatest energy. If the mass be magnetic, this line will stand axial; if diamagnetic, equatorial."

The remaining researches in Faraday's third volume are all of the highest interest and importance, most of them being connected with the subjects of magnetism and diamagnetism. The grand characteristics of all these researches are, great fertility of imagination, governed by admirable philosophic caution, and conscientious accuracy. In illustration of his deep insight, it may be said, that almost every result in diamagnetism has been anticipated by Faraday, though in affirmation he never allowed himself to take one step in advance of his experiments. Anxious only to interpret nature's laws faithfully, few philosophers have ever admitted their errors more frankly, and few have been able to obtrude themselves and their individual interests so little. In short, intellectually and morally, Faraday is a philosopher of the highest rank, of whom the century has just reason to be proud.

These reflections naturally remind us of the translation, under Colonel Sabine's superintendence, of Arago's valuable meteorological essays,[3] the noble introduction to which has been written by Arago's illustrious friend, Alexander von Humboldt. These essays, a great part of which refer to the phenomena of thunder and lightning, display great erudition and critical acumen. They are written in a lucid, popular manner, and will be found highly interesting to general as well as scientific readers. The same volume also contains essays on electro-magnetism, on animal electricity, on terrestrial magnetism – with copious notes by Colonel Sabine – and on the aurora borealis. Amongst the essays on electro-magnetism, the one on rotation magnetism will be read with an interest not unmixed with sadness. The phenomena of rotation magnetism, which were first effectually explained by Faraday in 1831, were discovered by Arago in 1824. Almost thirty years after his discovery, Arago, with shattered health and lost eyesight, conscious of his approaching departure from this world, reviews all that has been done and said upon the subject; and it is sad to find that the last hours of the venerable philosopher were embittered by the thought that posterity might not give *him* that credit in the discovery which was his right. Instead of enjoying that tranquillity of mind he had so nobly earned, we find him smarting under an old wound, and defending himself against an article in the "Edinburgh Review," which, in all probability, has long and deservedly been forgotten. Regretting, as we do, the injustice of that article, still we cannot but deem it far below Arago's notice. If there be any credit in being the first to observe a new fact, it is certainly small, and far below the credit of having ably, if unsuccessfully, endeavoured to discover its true cause. The greater credit is universally awarded to Arago; might he not safely have left the smaller credit, which was disputed only by a reviewer, in the hands of time?

The two volumes of his "Astronomie Populaire," which have appeared in France, have already been translated, under the superintendence of Admiral Smyth and Mr. Robert Grant, and published in one volume.[4] The great object of Arago, in his lectures at the Observatory at Paris, of which this book is a summary, was to make his favourite science understood by men unacquainted with physics or mathematics, and in this object he has succeeded admirably. Gifted with that rare talent of lucid exposition, and himself perfect master of all the details of his subject, he knew well what to withhold and what to impart in order to sustain the interest of his hearers or readers, and secure their permanent instruction.

We are glad to notice the appearance of a new edition of "Grove's Correlation of Physical Forces,"[5] for although experiment can scarcely be said to warrant some of its speculations, the book displays much acute thought. Mr. Wilson has favoured us with a still more speculative and metaphysical production.[6] His tract will interest many, but will assist but little the progress of science. Mr. Henfrey has done agriculturists a great service, which we hope they will appreciate, by his translation of Stockhardt's useful treatise on agricultural chemistry.[7] Mr. Hopkins has inflicted upon us a new edition of his absurd book.[8] It is not even worthy of criticism, and the fact of its having reached a third edition reflects anything but credit upon the scientific culture of our "enlightened British public." The Editor of the "Mechanics' Magazine" lowers the rank of his publication by admitting such trash as Mr. Harrington's "New Theories of Light and Heat."[9] A single extract taken at random from either of these two productions would sufficiently justify our sentence, and probably amuse the reader, but we regret the possible consequences of their publication far too much to encourage the same even by pronouncing them amusing. In one way only can such nonsense be beneficial: we may be induced thereby to encourage the publication of works containing sound elementary instruction. An opportunity for so doing is now afforded us by the appearance of Mr. Hughes's "Reading Lessons."[10] Mr. Hughes's object is a good one, and the energy with which he is realizing that object deserves praise. He has been one of the first to act upon the conviction, that only the ablest men can successfully teach the elements of their respective sciences. We do not say that his efforts have been crowned with complete success, but the list of writers whom he has engaged is sufficient guarantee for the general excellence of the lessons. Mr. Mann, a co-operator with Mr. Hughes, is himself the author of an excellent "Elementary Reading Book,"[11] which also deserves success. It is written in a lucid and attractive style, well suited to children.

The appearance of a second edition of Professor Rymer Jones's "General Outline of the Organization of the Animal Kingdom,"[12] with a large amount of additional matter, and with sixty new illustrations of the same character with those which constituted so remarkable a feature in the original work, demands from us more than a mere passing notice; since this treatise is the one to which those students almost necessarily have recourse, who desire to obtain a comprehensive and yet sufficiently-detailed account of the principal forms of animal structure. The first edition made its appearance in a complete form (having been issued in numbers) about fourteen years ago; and notwithstanding some blemishes, it acquired for itself a high and deserved reputation, as containing a large amount of accurate information, well selected and judiciously arranged, respecting the structure and physiology of each of the principal groups of which the animal kingdom consists; the descriptions being illustrated by a copious series of wood-engravings, such as had never been equalled in any work of its class. All these features of excellence are retained in the present edition, but the blemishes, we regret to say, are also more prominent; and whilst we can still recommend the work as containing a collection of clearly-imparted and reliable information, such as the student will not find elsewhere, it is our duty also to put him on his guard against certain errors and imperfections, by which he is liable to be grievously misled. In the interval which has elapsed since the appearance of the previous edition, a vast extension of our knowledge has been effected in regard to almost every class of invertebrated animals, rather, we regret to say, by the labours of Continental than by those of British naturalists; and the result of this extension has been to remodel, as it were, our conception of each group, necessitating, in many cases, an entire change in its zoological position. Further, the appearance of the admirable "Lehrbuch der vergleichenden Anatomie" of Von Siebold and Stannius, in 1848, which embodied a most complete and conscientiously-executed summary of everything which had been satisfactorily accomplished up to that date, afforded (as it were) a new starting-point, from which it might have been expected that any subsequent writer on comparative anatomy would have found it advantageous to take a fresh departure. For Professor Rymer Jones, however, these able authors, – of whom Professor Siebold is especially known by his numerous original contributions to various departments of his science, of a character which gives to his appreciation of the labours of others a value far above that of the mere compiler, – seem almost to have laboured in vain. Taking, as was not unnatural, his own first edition as his basis, he has seemed to cling to all its errors with a paternal fondness; the changes which have been introduced into the present edition being almost entirely in the way of *addition,* scarcely anything having been done in the way of *correction* or of *substitution.*

When, at the opening of the volume, we observed that Professor R. Jones had adopted from Siebold the group of *Protozoa,* and had associated under it, in what we hold to be their legitimate relation, the *Rhizopoda, Foraminifera, Sponges,* and *Infusoria,* we began to hope that we might find a like renovation to have been made in other parts of the work. But as early as the second page of this chapter, we are startled at the introduction of the *Spermatozoa* into the group; these being spoken of as "of kindred structure," notwithstanding that the history of their development (as cited by our author himself) clearly indicates that they belong to a category altogether different, being, in fact, no more entitled to the designation of independent Animalcules, than are blood-corpuscles or epithelium-cells. We could point out many other grievous errors of omission and of commission in this chapter, but must pass on to the two chapters that succeed it, which treat of the *Alcyonian* and *Actiniform* zoophytes; our author's account of these indicating the state of chaotic confusion wherein his mind seems to have remained, as regards this subject, from the time when he first wrote upon it. No one who has mastered the essential features of the organization of these two types of zoophytic structure, and has a proper appreciation of the value of characters, can have any difficulty in distinguishing what belong to each; and the separation has been most accurately made by Mr. Dana, the accomplished naturalist of the United States Exploring Expedition, who has probably a more profound acquaintance with the existing forms of the class than any one else can lay claim to. Yet all his labour has been thrown away upon Professor Rymer Jones, who jumbles together the forms belonging to these two very different types, as if no one had ever shown them to be distinct. Thus, he begins his chapter on *Anthozoa* with an account of *Fungia,* which he describes as it were but a little elevated above a Sponge, the "gelatinous investment" of its calcareous skeleton giving (according to him) but "dubious indications of vitality;" and in the same category he ranks the composite *Meandrinæ* with other *Madrephyllidæ.* Thence he proceeds to the *Alcyonian* zoophytes, of which he gives (after Milne Edwards) a sufficiently full and accurate account; but in immediate association with these we find the *Madreporidæ,* which are not only quite misplaced by being connected with them, but also, by being thrust-in here, separate the Alcyonians from the *Corallidæ, Pennatulidæ,* and *Tubiporidæ,* which belong to the Alcyonian group. In the succeeding chapter, the *Actinia* or sea-anemone stands all alone in its glory; and not a hint is anywhere given of the fact, that the previously-described Fungia, as well as the Madrephyllidæ and Madreporidæ generally, are all formed upon the same type; – a fact which is perfectly obvious in the case of the *Caryophyllia,* the little English madrepore, which any ordinary observer would take for a sea-anemone, its possession of a stony base of its own formation being its chief distinctive character.

In the next chapter, on *Hydrozoa,* we find a large amount of additional matter relating to the reproduction of that group, chiefly derived from the observations of Van Beneden; but the misapprehensions into which that eminent zoologist was led by theorizing where a gap remained in his facts, have been but very imperfectly set right by Professor R. Jones, who appears to be unacquainted with the admirable Memoir of Professor Allman on *Cordylophora,*[13] wherein the apparently diverse phenomena of medusiform gemmation are clearly and philosophically harmonized with each other. The change in the title of the succeeding chapter from *Acalephæ* to *Hydrozoa,* and the introduction of an account of the development of *Cyanæa aurita* (after Sars), show that the author is not ignorant of the close relationship existing between the Medusan Acalephæ and the Hydraform zoophytes; but there his knowledge seems to come to a dead stop; for not only does he take no notice whatever of the researches of Huxley, Leuckart, Kölliker, Vogt, and others, by which those perplexing groups, the *Physograda, Cirrigrada,* and *Diphyda,* have been (as it were) unravelled, but he repeats all his former misconceptions, apparently in the most blissful ignorance that any Naturalist has applied himself to the study of these animals since the date of his first edition.

To any one who has followed the progress of Helminthology during the last few years, it must seem strange to find the *parasitic worms,* which are clearly degraded members of the Articulated series, thrust in between the Hydrozoa (radiated animals) and the Bryozoa (molluscous), simply because of that rudimentary condition of their nervous system, which they share with the lower forms of each of the other types just named. In entering upon this chapter, we were startled at finding the order of *Cystic* Entozoa still retained, and supposed that by a want of attention to modern researches, corresponding to that which we have already had to note, our author had left himself unacquainted with those remarkable investigations, which now seem fully to justify the doctrine first advanced (we believe) by the sagacious Siebold, that the Cystic worms are but peculiar incomplete forms of the Cestoid, the differences of the two depending merely upon the locality wherein they are respectively developed. We find him subsequently, however, quoting largely from Van Beneden as to this point; but his knowledge seems to be entirely restricted to the researches of that observer, no notice whatever being taken of the labours of Siebold, Leuckart, Kuchenmeister, and others, who have contributed to elucidate it. And those who are acquainted with the extensive development which (as shown by Siebold) the "water-vascular system" presents in this group, will think it "passing strange" that no mention of it whatever is made by our author.

The *Bryozoa* still occupy, in Prof. Rymer Jones's view, the position he originally assigned to them, between the higher Entozoa and the Rotifera; notwithstanding that they bear not the slightest relationship to either one of these groups, and are now placed, by the almost unanimous voice of the most competent systematists, at the bottom of the Molluscous series, in close proximity with the Compound Tunicata. The transition from the Bryozoa to the Rotifera is still held by our author to be established through *Stephanoceros,* notwithstanding that it has been clearly shown to be a Rotifer in all the essential features of its structure, and to have nothing whatever in common with the Bryozoa, save a certain superficial analogy of conformation. In other respects, however, the chapter on *Rotifera* is considerably improved, due notice being taken of the observations of Dujardin, Dalrymple, Williamson, and Huxley: and it is in a citation from the last-named author, that we find the first and almost the only mention of the "water-vascular system" that the volume contains. From the Rotifera, we are carried onwards to the *Epizoa,* which are still treated as a separate class, notwithstanding that the tendency of all modern research has been to show, that they are nothing else than peculiar forms of Entomostracous Crustacea, whose females have acquired a strangely-aberrant form, consequent upon the excessive development of their egg-producing apparatus. That the males are conformable to the ordinary Entomostracous type, and that the females are so, too, at an early stage of their lives, seems to leave no doubt on this point; yet not the slightest indication of this relationship is given by Prof. R. Jones. It is through a like want of appreciation of the value of characters, that the *Cirrhopoda* are left in the place which they occupied in the former edition, namely, at the commencement of the Molluscous series; though every zoologist who bases his classification on organization and development, instead of being guided by superficial analogy, ranks them without hesitation in the Articulated series, in close proximity to the Crustacea, if not actually as a subdivision of that class. It would have been almost less discreditable to the author, to have ignored altogether the facts which modern investigations (especially the admirable researches of Mr. Darwin), have brought to bear upon this subject, than to have shown so extraordinary a want of appreciation of their value, as must utterly destroy any confidence that might have been otherwise placed in his guidance through the labyrinth of animal organization.

We had marked a great number of other points for comment; but we are constrained to remember that our duty keeps us within the limits of a notice, instead of permitting to us the discursiveness of a review. Having always highly approved the general plan of the book, we looked forward with hope to the time when the appearance of a new edition should give the author an opportunity of freeing it from its original defects, and of bringing it up to the present level of anatomical and zoological science. In this hope, however, we have been greatly disappointed; and we have discharged a painful duty in referring to the mistakes and shortcomings of a work which, with a little more care and discrimination, might have been made a most valuable compendium, supplying a widely-experienced want.

The curious discovery has been recently made, by Dr. Fred. Mone, in the library of the Benedictine Monastery of St. Paul, in Carinthia, of a palimpsest MS. of a portion of the "Natural History" of Pliny,[14] consisting of fragments of the 1st, 11th, 12th, 13th, 14th, and 15th books. This manuscript, having been liberally intrusted to the care of Dr. Mone by the Superior of the Monastery, he has applied himself to its decipherment – a matter of no small difficulty, since the writing was quite invisible until brought into view by chemical means; and having satisfied himself and others that it dates back as far as the fifth century, and that it is consequently the oldest known copy of any portion of Pliny's writings, he has prepared an exact reprint of it, page for page and line for line, with notes showing its departures from the currently-received text. This publication, which will doubtless be interesting to such as take a scholarly interest in the restoration of the text of an author of no mean repute, appears in a corresponding form with a very elegant reprint of the excellent edition of all the extant works of Pliny, by Gronovius, with additional notes, indices, &c., now nearly completed, which has issued from the Gotha Press.

One of the recent volumes of Mr. Bohn's serial issue contains the continuation of the translation of Pliny's "Natural History,"[15] executed by the late Dr. Bostock and Mr. Riley. We are glad that Mr. Bohn has included this in his Classical rather than in his Scientific Library; for whatever may be its interest to the classical scholar or to the antiquarian, the naturalist most assuredly never has recourse to it as a scientific authority, so strangely indiscriminate is the mixture it contains of truth and error, of information that would be most valuable if reliance could be placed upon it, with "old wives' fables." In this respect the writings of Pliny contrast most unfavourably with those of Aristotle; the truthfulness of whose statements respecting the structure and habits of the marine animals of the

Ægean, becomes more and more apparent the more they are investigated; whilst his notions of classification, based on internal organization, were far sounder, not only than those of his contemporaries, but even than those of most of his successors down to a comparatively recent period.

In the new edition of Mr. Currey's translation of Dr. Schacht's "Treatise on the Microscope in its Application to Vegetable Anatomy and Physiology,"[16] the translator has introduced some additional chapters on the forms of microscope and on the accessory apparatus commonly used in this country; and he has, besides, been furnished by Dr. Schacht with much new matter, embodying the results of his investigations since the publication of the original work three years since. At the suggestion of Dr. Schacht, also, he has added three chapters on the embryogeny of the Coniferæ, from that author's larger "Beitrage zur Anatomie und Physiologie der Gewächse." These additions greatly enhance the value of this little treatise to the English student; and we hope that it will tend to promote the prosecution of original investigations in this department of science, in which the botanists of our own country seem at present to be content to allow all the honours to be carried off by their Continental brethren. In our recommendation of the book, however, we must insert a caution as to the implicit reception of the author's statements with respect to the fertilization of the ovule; for on this point he follows Prof. Schleiden, and is at issue with Hoffmeister, Mohl, Tulasne, Henfrey, and other excellent observers. And although he has recently adduced what he considers to be a new and triumphant verification of Schleiden's doctrine that the embryo is developed within the extremity of the pollen tube, yet his opponents do not admit that the preparation shows what he affirms it to do. Those who are interested in this discussion, will find some of the latest news of it in the "Annales des Sciences" (fourth series), Tom. iii., No. 4.

Dr. Mc Cormac's treatise "On the Nature, Treatment, and Prevention of Pulmonary Consumption, with a Demonstration of the Cause of the Disease,"[17] is a phenomenon in medical literature. That a young man desirous of puffing himself into notice by means not the most legitimate, should publish such a flimsily written and pretentious little book, would not so much surprise us. But that a physician of large experience and extensive practical information, of various reading and of highly cultivated mind, as Dr. McCormac has shown himself on former occasions to be, should risk his reputation by putting forth the most unfounded claims to a discovery which is no discovery at all, because it is palpably not true, astonishes us more than we can express. The whole gist of his doctrine is put forth in his preface, which is as follows: –

> "Consumption and scrofula, in all essentials, are one; tubercle, in its varied protean guises, is but the result of a deterioration of the blood, of the retention of excretions, carbonaceous, and other impurities in the blood, where they have no business to remain. In consequence of the imperfect performance of the respiratory functions, these impurities accumulate. The time at length comes round when they must be got rid of; if not

> during and through, the act of respiration, perforce otherwise. The result is their deposition, as an inorganic matter, in the form of tubercles in the lungs, and other organs, these, saving the diffusion of tuberculous blood, being perchance, in other respects, sound. A dead matter is deposited in the tissues; that which should be thrust aside – shovelled out, as it were – is detained within the organization. These all-important facts *being positively determined,* the cause and radical cure – in short, the prevention of consumption and scrofula, with all their concomitant ravages, are placed as absolutely as small-pox itself has been placed, within human control. There need now be no more consumption, no more scrofula; and diseases which have actually advanced as civilization itself has advanced, henceforth, now and for ever, may be set aside."

Now Dr. McCormac's fundamental proposition is essentially untrue. Tubercular matter is not an impurity resulting from the retention of excrementitious products, but is nutritive material which has not been brought, by the assimilating process, up to the requisite standard of organizability for being applied to the repair of the tissues, and is therefore deposited in a state which departs more or less (in the different forms of tubercle) from the structure of the parts it ought to nourish. Hence the aim of the rational physician is not to "shovel" tubercular matter out of the body, but to render it more organizable; that is, more fit to be converted into living tissue; and it seems to be by its influence in promoting the deficient assimilation, that cod-liver oil exerts that beneficial influence, about which few physicians of large experience entertain any scepticism. Dr. McCormac's one idea is, that deficient respiration is the essential cause of tubercular deposit, and therefore that consumption and scrofula may be prevented by the promotion of free and complete aëration of the blood. He adduces a number of facts, with which every well-informed medical man is familiar, with reference to the share which imperfect aëration of the blood has in the production of tubercular disease; but he does not advert to a number of others, which prove that it is by no means the sole cause. We will give him one, which he seems altogether to have overlooked. According to the very trustworthy report of Dr. Schleusner, who was sent by the Danish government a few years since to investigate the sanitary condition of Iceland, no combination of what are commonly accounted the predisposing causes of consumption and scrofula could be more complete, than that which exists among the mass of the Icelandic peasantry. Whole families are huddled up with their sheep, not only during the night, but during the greater part of the day, for half the year, in most miserable hovels, destitute of any ventilation but that afforded by the chimney; their clothing is not once put off or changed during the whole of that time; their food is scanty; and the external atmosphere is both cold and damp. The unhealthy condition of the population is evidenced by its extraordinary liability to epidemic disorders, and by its want of increase, or even (in some districts) by its absolute diminution. And yet amongst this remarkable people – the best educated peasantry in Europe, so far as regards what is commonly accounted education – scrofula and consumption are unknown. The only ostensible cause for this immunity lies in the highly oleaginous nature of their diet, which consists in great part of the oily bodies of piscivorous birds. To affirm, then, that deficient respiration is, *per se,* the cause of tuberculous diseases, and that due attention to the vigorous performance of this function is sufficient to prevent, or even to cure them, is most assuredly to go far beyond the warranty of facts; and the only difference between Dr. McCormac's system and that of the consumption-curing quacks, lies in this, that his method and aims are not only good as far as they go, but, when put in practice as preventive measures, can do nothing but good; whilst those of the gentry to whom we allude, often do a great deal of harm. We are as far as possible from desiring to discourage or to undervalue Dr. McCormac's efforts to awaken public attention to the vast importance of fresh air; but we are anxious to make it understood that no amount of fresh air will of itself prevent, still less cure, consumption, though the free use of it would doubtless greatly diminish the ravages of that fell destroyer.

We have received from the other side of the Atlantic a "Monograph of Mental Unsoundness,"[18] by Francis Wharton, which, we are informed, is intended to constitute the first part of a general treatise on medical jurisprudence, to be produced by its writer in conjunction, with Dr. Moreton Stille, of Philadelphia. We infer from the whole tenor of this production, that it is the work of a lawyer, who has been accustomed to view insanity merely in its juridical relations, and not of a psychological physician, who has made a study of the essential nature of the disease, and of the varied forms under which it manifests itself. It contains a great mass of facts, drawn from the practice of the courts on both sides of the Atlantic; but this serves little other purpose than to show the utter want of any definite apprehension in the legal mind as to what really constitutes insanity, and the consequent inconsistencies in the decisions which are referred-to as precedents. For example, we cannot imagine any unprejudiced man of common sense, much less a medical practitioner, to study the evidence given on the recent trial of Buranelli, and the additional facts subsequently brought before the Home Office, without coming to the conclusion that he was insane enough to be irresponsible for his actions, and that his execution was consequently a judicial murder; whilst, on the other hand, many a prisoner has escaped punishment on the plea of insanity, who has had no more claim to the exemption than might be advanced for a man who should knowingly drink himself into a fit of passion, and should then murder his wife or commit some other outrageous crime. So long as the existence of delusion is held to be the chief test of insanity, and the knowledge of right and wrong to be the sole test of criminality as opposed to irresponsibility, so long will the *principle* of our legal decisions be radically wrong. The whole mistake has arisen from the ordinary limitation of the view to fully-developed cases of insanity, of which delusions form a common, though not by any means an uniform feature. For the more carefully the disease is studied through its whole course, the more does it become apparent that in by far the larger proportion of cases, morbid impulses, emotions, and feelings, and the loss of volitional control over them, constitute its most essential and constant feature; whilst delusions of the intellect are only developed during the progress of the disease, and frequently do not present themselves at all. Thus, so far from what has been designated as "moral insanity" being the *exceptional,* it is really the *ordinary* form of the disease; and the question is not, what is the false belief, but what is the perverted emotional state leading to the production of that belief, and how far are the actions to which it prompts under the control of the rational will. We do not pretend to offer any new test of insanity; every case must be judged by its own merits, and ought to be decided, not by an ignorant jury under the guidance of legal prejudices, but by a tribunal of medical men who have made insanity their special study. Reference of all doubtful cases to such a tribunal appears to us to be as obviously the course which the circumstances of the case require, as is the reference of a knotty point of law to the opinion of the highest legal authorities; and we no more admit the competency of a lawyer to sit in judgment upon a question of psychological medicine, than we should uphold the competency of a court of doctors to lay down a precedent in a case involving the nicest judicial questions. At present, as our readers are doubtless aware, medical men are only examined as witnesses in cases of this kind; and the weight attached to their evidence, indeed the very permission to deliver an opinion, depends mainly on the inclination of the presiding judge; so that it has not unfrequently happened that the men most competent to guide the decision of the jury, have been virtually put out of court. We can only look to the influence of an intelligent public opinion as a means of remodelling the present most unsatisfactory system; and that influence we are most desirous of bringing to bear upon the question, whenever the time shall arrive for its being exerted to any good purpose.

From Sir H. Holland we have a new edition of his "Medical Notes and Reflections;"[19] a work which has done much to improve the general tone of thought upon medical subjects, in the mind both of the profession and the public. It consists of a series of detached essays upon various points both of theory and practice, in which the object kept in view is not so much to arrive at dogmatic conclusions, as to exhibit the mode in which they should be sought-for by minds imbued with philosophic habits of thought; and we know of no guidance to which we could more unreservedly commit the formation of such habits than that of the accomplished author of the work before us. The previous editions contained a few chapters on Psychological subjects, especially directed to the elucidation of those mutual actions and relations of bodily and mental phenomena which enter into the life of man. These chapters were detached by the author about three years ago, and were embodied with much new matter in a separate volume, under the title of "Chapters on Mental Physiology," with the view of giving a philosophical direction to the public mind on the interpretation of the phenomena of electro-biology, table-turning, &c., which then occupied its attention. And if they in any degree fall short of this object, it was simply because they were too highly pitched for popular apprehension, their cautious and suggestive tone being fitted rather to impress itself upon the thinking few, than upon the unreasoning mass, which looks for confident assertion and ready-made conclusions. These chapters have been replaced in the present volume by some additional essays of the same general character with those which constitute the bulk of the work; and instead of being thrown together without arrangement, the chapters are now grouped in an appropriate order, so as to constitute three very natural series – the first on general questions of medical philosophy, the second on special forms of diseased action, and the third on the operation of particular classes of medicines. In the discussion of each of these subjects, we find the author constantly endeavouring to "separate what is actual and assured knowledge from that which but usurps the name and show of it," and "to place in the clearest form those principles of thought, observation, and conduct, which may most conduce to the progress of medicine, and the honour and usefulness of those who profess it." And nothing can give a juster title to offer such advice, than the experience of a mind so richly stored with ancient and modern learning, so wisely governed and so loftily directed, in forty years of active and successful practice. If the rising generation of medical men would "read, mark, learn, and inwardly digest" the wise counsels of this volume, the public would have no reason to complain of those unphilosophical habits of reasoning and injurious methods of practice, which have alienated so many from legitimate medicine, and have driven them into the embrace of the various quackeries which succeed one another from time to time in popular favour.

The title of Dr. Lindley Kemp's "Phasis of Matter"[20] led us to expect a performance of a kind very different from that which the author has attempted. We had supposed that he would endeavour to exhibit the succession of varied aspects under which the same particles of matter may present themselves, in the solid mass of the earth, diffused in the gaseous constituents of the atmosphere, again consolidated by the agency of the plant into those organic compounds that furnish the mechanism of life, transferred to the animal body, in which, they become the material instruments of thought and feeling, of intelligence and will, and then again resolving themselves into the dust from which they were at first drawn. For the worthy treatment of such a subject, an author would require a large and accurate acquaintance with geology and physics, chemistry and physiology; together with a power of seizing their essential points of connexion, and of giving somewhat of a pictorial grouping to the facts thus brought together. Dr. Kemp has restricted himself, however, to the more modest aim of giving "an outline of the discoveries and applications of modern chemistry," "intended for the wants of the general scholar and of men of the world, whose active occupations are more or less based upon a knowledge of chemical principles and chemical facts." The work is divided into four books, of which the first contains an outline of inorganic chemistry, as developed by the researches of the laboratory upon the properties of the elementary bodies; the second treats of the chemical nature of the compounds naturally existing in the minerals and rocks of the globe; the third describes the chemistry of the compounds peculiar to animal and vegetable organisms; and the fourth attempts to present an outline of the changes which matter undergoes in these bodies during their living state. On the whole, Dr. Kemp may be said to have succeeded fairly in his object. His book is a respectable performance, free from any glaring defects; but we cannot point to any feature of high excellence; and it seems to us to be rather the work of an author who has brought together his knowledge by reading alone, than of a really scientific thinker, who, having acquired a personal familiarity with the sciences of which he treats, has had something to tell of the harmonious and mutually-related aspect under which they present themselves to his own mind. – There are doubtless many readers to whom the ordinary text-books of chemistry present a very repulsive aspect, who will find in Dr. Kemp's volumes a readable summary of its more important facts and doctrines.

Source Endnotes

1 "Faraday's Experimental Researches in Electricity." Vol. III. London: Taylor and Francis, Red Lion Court, Fleet-street, 1855.
2 "Bakerian Lecture," 1855; and "Philosophical Magazine," vol. 10.
3 "Meteorological Essays." By Francis Arago. London: Longman, Brown, Green, and Longmans. 1855.
4 "Popular Astronomy." By Francis Arago. Vol. 1. London: Longman, Brown, Green, and Longmans. 1855.
5 "The Correlation of Physical Forces." By W. R. Grove. Third Edition. London: Longman, Brown, Green, and Longmans. 1855.
6 "The Unity of Matter." A Dialogue. By Alex. Stephen Wilson. London: S. Highley. 1855.
7 "Chemical Field Lectures: A Familiar Exposition of the Chemistry of Agriculture, addressed to Farmers." By Dr. J. A Stockhardt. Edited, with notes, by Arthur Henfrey. London: H. G. Bohn.
8 "An Illustrated Introduction to the Connexion of Geology and Magnetism." By Evan Hopkins. London: Taylor and Francis. 1855.
9 "Mechanics' Magazine." Numbers 1678, 1679, and 1680.
10 "Hughes's Heading Lessons." First and Second Books. London: Longman, Brown, Green, and Longmans. 1855.
11 "Lessons on General Knowledge." By Robert James Mann. London: Longman, Brown, Green, and Longmans. 1855.
12 "General Outline of the Organization of the Animal Kingdom, and Manual of Comparative Anatomy." By Thomas Rymer Jones, F.R.S., Professor of Comparative Anatomy in King's College, London, &c &c. Second Edition. Illustrated by Four Hundred Engravings. London: Van Voorst. 1855.
13 "Philosophical Transactions." 1853.
14 "C. Plinii Secundi Naturæ Historiarum." Lib. I. XI. XII. XIII. XIV. XV. "Fragmenta, e Codice Rescripto Bibliothecæ Monasterii ad S. Paulum in Carinthia." Edidit Fridegariua Mone. Ph.D. Gothæ: Andr. Perthes. 1855.

"Frederici Gronovi in aliquot C. Plinii Naturalis Historiæ Libros Notæ, emendatius editæ et locupletioribus indicibus auctæ." Gothæ: Andr Perthea. 1855.
15 "The Natural History of Pliny." Translated, with copious Notes and Illustrations. By the late John Bostock, M.D., F.R.S., and H. T. Riley, B.A. Vol. III. London: Bohn. 1855.
16 "The Microscope, and its Application to Vegetable Anatomy and Physiology." By Dr. Hermann Schacht. Edited by Fred. Currey, M.A. Second Edition, considerably enlarged, with numerous Illustrations. London: Highley. 1855.

17 "On the Nature, Treatment, and Prevention of Pulmonary Consumption, and incidentally of Scrofula, with a Demonstration of the Cause of the Disease." By Henry McCormac, M.D. London: Longmans. 1855.
18 "A Monograph of Mental Unsoundness." By Francis Wharton. Philadelphia: Kay and Brothers. 1855.
19 "Medical Notes and Reflections." By Sir H. Holland, Bart., M.D., F.R.S., &c. &c. Third Edition. London: Longmans. 1855.
20 "The Phasis of Matter; being an Outline of the Discoveries and Applications of Modern Chemistry." By T. Lindley Kemp, M.D. In Two Volumes. London: Longmans. 1855.

Editorial Headnote

9. "Electricity and the Future," *ST. James Magazine*, January 1882, pp. 97–102.

First published in 1861, the *St. James Magazine* sought to "Promote the Interests of Home, the Refinements of Life, and the Amusement and Information of all Classes." This is not to be confused with a similar British publication, also called the *James Magazine*, first published in 1763 by Robert Lloyd, which primarily contained literary pieces. The former magazine contained articles and reviews aimed at the general interests of Londoners, particularly the gentlemen who circulated in the clubs of the popular St. James Parish in London. The author of this essay, Frederick Charles Webb (1828–1899) was a civil engineer who worked on the laying of undersea cables, lighthouses, and dock illumination. Webb's essay makes clear that the electrical imaginary remained strong throughout the nineteenth century, and even among the people that worked most closely with electricity, boundless possibilities remained. Specifically, his essay touches on the *Exposition Internationale d'Electricité*, which famously brought some 1,758 inventions to the public eye and made names like Siemens, Brush, Crompton, Edison, and Jablochkoff household names. He explores the benefits of the dynamo electric machines and *magneto-electric machines* that could produce more electricity over greater distances by harnessing the force of water or steam power. In the sections he foreshadows the transformation of electricity into a commodity. He continues to discuss other major changes to telegraphic and telephonic communication. But he also points to a failing, a London failing. He notes that electric power and communication offer the opportunity for precision and control and that cities like Ostend can have electricity. In a nod to the magazine's primary readers, he asks why can't London? He concludes by calling for a focus on transforming the excitement around the Paris Exhibition into a practical use and for it to be put on display in the Crystal Palace in the immediate future.

9

ELECTRICITY AND THE FUTURE

F. C. Webb

Source: F. C. Webb, "Electricity and the Future," *ST. James Magazine*, January 1882, pp. 97–102.

THE brilliant success of the *Exposition Internationale d'Electricité* at Paris, has shown thousands the realisation of numerous applications of electricity which, though known to the scientific and the engineering world, were still only half believed in by those who have but read short paragraphs in the daily press relating to the various inventions which have been crowding forward within the last three or four years. The telephone and its great first lieutenant, the microphone, and all their various modifications of telephonia are, in spite of the practical applications they have received, still unknown to thousands of persons in those countries of Europe which are at any distance from England, France, or the United States.

The electric light, too, in spite of the practical illustrations, in London, by the Sociètie Generale d'Electricité with the Jablochkoff candle, the Anglo-American Brush Electric Light Corporation, Messrs. Siemens Brothers, The British Electric Light Company, Mr. Crompton, Mr. Maxim, and numerous others, still gained in publicity by the Paris Exhibition. But this is only writing of the most general and important applications of electricity. Many of those who glance over the carefully prepared catalogue of the Paris Exhibition will be astonished to find, in its two hundred carefully prepared pages, 1,758 different exhibits of electrical apparatus, without counting the list of rooms, imitation theatres, &c., showing forth the various modes of electric lighting, both by the arc and the incandescent systems. If Faraday could to-day for a moment return amongst us and survey these practical results of his patient researches after truth, he would at last be rewarded for his brilliant discoveries. The philosopher who, with a small piece of soft iron and a few yards of copper wire obtained the first symptoms of electricity from magnetism, could not but feel the most intense pleasure at the results which have grown from his systematic search to obtain an electric current from magnetism. Let us take a glance back at magneto electricity.

Faraday, seeing that since from an electric current magnetism could be produced (that is to say that by passing an electric current through an insulated wire wound round a piece of soft iron, the iron became magnetised) argued that there must be a means of obtaining an electric current in some way from a magnet. By careful and continual experiment, he at last found that by *moving* a coil of

DOI: 10.4324/9781003295556-11

insulated wire from or to a magnet, a current was produced in the coil whilst so moving; the current, when the coil was moving towards the magnet, being in the reverse direction to that obtained when the coil was receding from the magnet. This formed the basis of magneto electricity. Numerous inventors soon applied this discovery, and for years this mode of producing currents of electricity of short duration was used for various telegraphic instruments, the instantaneous currents of electricity thus produced being utilized to move step by step telegraph instruments.

Next came the employment of a commutator, which converts the alternating currents thus produced into a series of currents in the external circuit in *the same direction,* thus producing a continuous current of electricity.

Professor Holmes perfected a large machine for the Trinity Board of London, by which a series of coils revolving before permanent magnets, and with a proper commutator, gave out the alternate currents produced by the machine in one continuous direction, and this current was used to produce an electric light between carbon points.

Soon after came the great step in electricity, produced by movement of a coil of wire. It was argued at about the same time by Dr. C. W. Siemens, Mr. A. Varley, and Professor Wheatstone respectively, that, as the movement of a coil of wire before a permanent magnet would produce a current in the wire, it was first also possible to obtain the same current if the magnet was an electro-magnet (that is a magnet produced by a current of electricity passing through an insulated wire wound round a piece of soft iron). Secondly, there appeared no reason to these gentlemen why the current to excite the electro magnet should not, instead of being obtained from a battery, be obtained by the *current itself* produced in the revolving coils. But the difficulty at first appeared that, as the electro magnets were formed only of soft iron, there must be some means of exciting these magnets to start with.

Accordingly in the early ideas on this subject a few elements of a battery were to be used to excite the electro magnets (commonly called the field magnets). Then as the coils revolved before them and produced a current in the coils, this current was led round the field magnets, and exciting them to a higher degree, made of these strong electro-magnets. Then the coils revolving before these stronger magnets produced a greater current, and thus there was an action and re-action, which was only limited by the force available for driving the coils, for, as the magnetism in the field magnets increases, so does the resistance to the movement of the coils.

I have thus traced the action of a Dynamo-Electric Machine, one of the marvels of the present age, by which enormous streams of electricity are provided for the production, not of one electric light only, but of thirty or forty in one circuit.

This has been, of course, the work of many experimenters whom it would be impossible to name in the space of this article. I have only attempted to mention the transition period from those machines in which electricity was produced by moving coils of insulated wire near permanent magnets, or what were termed *magneto-electric* machines, to those machines which, employing a portion of the current generated by their own action, excited the field magnets before which the coils revolved, and which have been termed, from the fact that electricity is produced simply by *movement,* DYNAMO-ELECTRIC machines.

But it is necessary here to point out the enormous currents that are produced by these machines, as compared with ordinary batteries. Let us, for instance, calculate what number of Daniel elements (a class of element largely used, though in various forms, for telegraphic purposes) would be required to produce a similar current to that obtained from a Brush machine of the class specified by the Anglo-Ameriean Brush Electric Light Corporation as their 7A machine.

Taking the figures given by Mr. Brush, the designer of this machine, it is stated, in the number of "Engineering" for February 4, 1881, that the electro-motive force is equal to 839 volts. Now, a volt is, roughly speaking, equal to about the electro-motive force of an ordinary Daniels element. Consequently to have the electro-motive force of this machine we should have to employ 839 Daniels elements, which alone would be a formidable battery, since even the single coils of gutta percha covered wire used for submarine telegraphy are never tested with more electric power than 500 such elements. But this is not all, we have to consider the resistance of the battery. Now, the resistance of an ordinary Daniel's element may be taken at about 5 ohms (the accepted unit of resistance). Consequently, the resistance of such a battery would be five times 839 elements, or 4195 ohms. But the resistance of the No. 7 Brush machine is said to be, by Mr. Brush, 10.55 ohms. Consequently, to obtain a similar current from Daniel elements, we must place in parallel circuit such a number of batteries of 839 elements as will reduce the whole resistance of the combined batteries to 10.55 ohms. To avoid decimals in calculation, let us take Mr. Brush's figures for resistance of the 7A machine at 10 ohms, and then it will be seen, by Ohm's law, that to obtain an electro-motive force of 839 volts, with the resistance in the battery of only 10 ohms, we must divide 4195 by 10, giving 419.5 as the number of batteries of 839 Daniel's elements each, that must be joined up in parallel circuit to give a current equal to a 7A Brush machine.

The conversion of mechanical force into electric current thus effected in a dynamo machine has its converse, since the current produced (less the loss by dissipation of energy by re-distribution of heat) can be conveyed to a distance. For instance, a current sent from one dynamo machine can be made to move, so as to drive machinery of another at a distance. Thus we have Dr. Siemen's electric railway, and it is evident that many sources of power may be utilized similarly. Waterfalls may be made to turn turbines, which, in their turn, can drive dynamo machines, which, producing the powerful currents we have endeavoured to estimate, can, by suitable conductors, be made to produce – and, indeed, have already produced – electric lights at a distant point. Again, if even steam power is employed, by the means of central stations, with a proper arrangement of dynamo machines, currents can be (and are already in America) sent to any factory or private house where a current of electricity is required for electric light purposes, or for driving light machinery. Indeed, it is astonishing that amongst the enterprising men who abound in England, with unlimited professional advice, and energetic contractors, no company has been as yet started to establish central stations for supplying electric currents for illuminating purposes, and light mechanical work. The contracting and manufacturing firms cannot be complained of as regards extensive exhibits, but it is clear that such firms must have some limit to their expenditure. Manufacturers of railway plant are not expected to go alone to Parliament for powers to construct railways and work them, and it is certainly asking too much of our manufacturers of electric apparatus to leave to them the distribution of energy by means of central stations. Companies should, and no doubt will in the immediate future, be formed to provide this in every town. In England, where most towns are already lit by gas, the gas interest has to be fought, but in many countries such as, for instance, Portugal, Spain, Italy, Greece, Algeria, Norway and Sweden, there are many hundreds of towns of considerable size where gas has not yet appeared, and the municipalities are eager to rise from a wretched system of oil lights to the use of electricity. The small incandescent lights must be offered them for their towns, theatres and shops, as well as the arc lights for their streets, squares and railway stations. The future is, in every sense of the word, brilliant, if only enterprising men come forward to step between the manufacturer and the public.

Turning now to the Telephone of Bell, Hughes and Edison, the beautiful illustration of its use given at the Paris Exhibition, where opera and comedy could be distinctly heard without the trouble of crushing to the house, inspires us with the hope that stations may soon be established in London, when, by the payment of a shilling, we may be able to hear our particular air or duet in “Les Hugenots” or “Faust” without sacrificing the whole evening to the tyranny of a heated theatre.

Of the numerous improvements in telegraphy which the Paris Exhibition has brought to the front, it would be impossible within the limits of this article to speak. Duplex, and steam quadruplex, and Edison and Wheatstone automatic may soon be out of date, as improvement after improvement comes to the front. The number of pages in the catalogue of the Paris Exhibition testifies to this.

Amongst the many applications of electricity, there is one in which we in London are singularly behind our time. In many towns on the continent clocks are established by the municipalities at the corner of nearly every street, worked by electric currents from a central and well regulated clock. No town requires such clocks more than London, where time is money.

Surely if such places as Ostend can have a system of electric clocks, London ought to have one, and it is a disgrace to our municipal or divided parish system of Government that we should be dependent on either private watches or on a stray church clock, which, if not wrong, is generally stopped.

Again, as regards what may be termed short telegraphy, the conveyance of signals to the helm and engine-room on board ship, adopted considerably in telegraph repairing ships (notably in those of the India Rubber, Gutta Percha, and Telegraph Works Company) opens up an ample field for the inventor if supported by enterprising capitalists.

In the same way our dwelling-houses could be supplied with the means of asking for hot water, coals, plates, or a thousand things, without having to bring an unfortunate servant up endless stairs.

There seems, in fact, no limit to the advantages and benefits which electricity and its now well-known laws can confer on us in future, but we must have more support from enterprising commercial men. Electricians seem to be struggling too much by themselves, both in electric lighting and the few other branches I have endeavoured to point out. Telegraphy, no doubt, is on a firm basis, supported strongly by capital, but there is ample room for improvement in numerous other branches of electricity, which it may be hoped the forthcoming exhibition at the Crystal Palace will bring into practical use in the immediate future.

F. C. Webb, C.E.

Editorial Headnote

10. "Electricity in Relationship to Science," *Popular Science Monthly* 40, February 1892.

The British scientist, experimenter, and public lecturer William Crookes (1832–1919) worked as a chemist and physicist in the Royal College of Chemistry, now part of the Imperial College in London. Crookes circulated through many of the most well-known institutions of science of the time throughout his life, including the Royal Institution, British Association for the Advancement of Science, Royal Society, and American Philosophical Society. He also worked closely with scientists like Michael Faraday (1791–1867), J. J. Thompson (1856–1940), George Gabriel Stokes (1819–1903), and gained notoriety for working closely with notable spiritualists like Kate Fox (1837–1892) and Daniel Home (1833–1886). He is probably best remembered for his development of the Crookes Tube, spectroscopy, and discovery of thallium. In addition to his more specialized research, Crookes developed a great deal of fame for his work on public demonstration and public communication of science. It is in that category the publication shown here fits. The American magazine *Popular Science Monthly* began circulation in 1872 under the editorship of Edward L. Youmans (1821–1887) and continues today as a magazine containing written and illustrated content for the "educated layman." British authors and scientists frequently wrote or were featured in its pages. In this particular issue, a speech by Crookes which also featured in *Nature* was published, demonstrating the trans-Atlantic captivation with the possibilities and realities of electric power and associated technology. Crookes delivered his speech to the Institution of Electrical Engineers at the Criterion Theatre, a relatively recently refurbished and electrically illuminated theatre in London's West End. Crookes begins his speech by pontificating on the general benefits of all experimentation, "irrespective of practical applications," by nodding to the experimentation of Galvani on frogs which led to the telegraph and telephone. He uses this to comment on the possibilities of understanding the "chemical atom" or unlocking the potential use of hitherto-untapped energy. Also, he makes note of the ongoing alternating current debate by citing Tesla and the negative reputation. Some of the most jarring suggestions of the speech include his arguing that coal may soon be irrelevant and his hopes to resurrect the scientific study of the relationship between electricity and life. Much as with Webb's article discussed earlier, Crookes makes clear that even at the end of the nineteenth century, for the most highly trained scientific minds and the public electricity and its potential seemed limitless and occupied a vast imaginary in which it functioned as a salve for old issues and for problems that did not yet exist. For Crookes, and many other authors at the time, it is worth noting that electricity occupied a space as a science but also as a tool to enhance other areas as people wrestled with new ways to understand nature.

10

ELECTRICITY IN RELATIONSHIP TO SCIENCE[1]

William Crookes

Source: William Crookes, "Electricity in Relationship to Science," *Popular Science Monthly* 40, February 1892.

THE third annual dinner of the Institution of Electrical Engineers was held at the Criterion on Friday, November 13th, Prof. William Crookes, the president, was in the chair. In proposing the toast of the evening, "Electricity in relation to Science," Prof. Crookes delivered the following speech:

We have happily outgrown the preposterous notion that research in any department of science is mere waste of time. It is now generally admitted that pure science, irrespective of practical applications, benefits both the investigator himself and greatly enriches the community. "It blesseth him that gives, and him that takes." Between the frog's leg quivering on Galvani's work-table and the successful telegraph or telephone there exists a direct filiation. Without the one we could not have the other.

We know little as yet concerning the mighty agency of electricity. "Substantialists" tell us it is a kind of matter. Others view it, not as matter, but as a form of energy. Others, again, reject both these views. Prof. Lodge considers it "a form, or rather a mode of manifestation, of the ether." Prof. Nikola Tesla demurs to the view of Prof. Lodge, but thinks that "nothing stands in the way of our calling electricity ether associated with matter, or bound ether." Higher authorities can not even yet agree whether we have one electricity or two opposite electricities. The only way to tackle the difficulty is to persevere in experiment and observation. If we never learn what electricity is, if, like life or like matter, it should remain an unknown quantity, we shall assuredly discover more about its attributes and its functions.

The light which the study of electricity throws upon a variety of chemical phenomena – witnessed alike in our little laboratories and in the vast laboratories of the earth and the sun – can not be overlooked. The old electro-chemical theory of Berzelius is superseded, and a new and wider theory is opening out. The facts of electrolysis are by no means either completely detected or coordinated. They point to the great probability that electricity is atomic, that an electrical atom is as definite a quantity as a chemical atom. The electrical attraction between two

DOI: 10.4324/9781003295556-12

chemical atoms being a trillion times greater than gravitational attraction is probably the force with which chemistry is most deeply concerned.

It has been computed that in a single cubic foot of the ether which fills all space there are locked up ten thousand foot-tons of energy which have hitherto escaped notice. To unlock this boundless store and subdue it to the service of man is a task which awaits the electrician of the future. The latest researches give well-founded hopes that this vast storehouse of power is not hopelessly inaccessible. Up to the present time we have been acquainted with only a very narrow range of ethereal vibrations, from extreme red on the one side to ultra-violet on the other – say from three ten-millionths of a millimetre to eight ten-millionths of a millimetre. Within this comparatively limited range of ethereal vibrations, and the equally narrow range of sound vibrations, we have been hitherto limited to receive and communicate all the knowledge which we share with other rational beings. "Whether vibrations of the ether, slower than those which affect us as light, may not be constantly at work around us, we have until lately never seriously inquired. But the researches of Lodge in England, and Hertz in Germany, give us an almost infinite range of ethereal vibrations or electrical rays, from wave-lengths of thousands of miles down to a few feet. Here is unfolded to us a new and astonishing universe – one which it is hard to conceive should be powerless to transmit and impart intelligence.

Experimentalists are reducing the wave-lengths of the electrical rays. With every diminution in size of the apparatus the wave-lengths get shorter, and could we construct Leyden jars of molecular dimensions the rays might fall within the narrow limits of visibility. We do not yet know how the molecule could be got to act as a Leyden jar; yet it is not improbable that the discontinuous phosphorescent light emitted from certain of the rare earths, when excited by a high-tension current in a high vacuum, is really an artificial production of these electrical rays, sufficiently short to affect our organs of sight. If such a light could be produced more easily and more regularly, it would be far more economical than light from a flame or from the arc, as very little of the energy in play is expended in the form of heat-rays. Of such production of light. Nature supplies us with examples in the glow-worm and the fire-flies. Their light, though sufficiently energetic to be seen at a considerable distance, is accompanied by no liberation of heat capable of detection by our most delicate instruments.

By means of currents alternating with very high frequency. Prof. Nikola Tesla has succeeded in passing by induction through the glass of a lamp energy sufficient to keep a filament in a state of incandescence without the use of connecting wires. He has even lighted a room by producing in it such a condition that an illuminating appliance may be placed anywhere and lighted without being electrically connected with anything. He has produced the required condition by creating in the room a powerful electrostatic field alternating very rapidly. He suspends two sheets of metal, each connected with one of the terminals of the coil. If an exhausted tube is carried anywhere between these sheets, or placed anywhere, it remains always luminous.

The extent to which this method of illumination may be practically available experiments alone can decide. In any case, our insight into the possibilities of static electricity has been extended, and the ordinary electric machine will cease to be regarded as a mere toy.

Alternating currents have at the best a rather doubtful reputation. But it follows from Tesla's researches that as the rapidity of the alternation increases they become not more dangerous but less so. It further appears that a true flame can now be produced without chemical aid – a flame which yields light and heat without the consumption of material and without any chemical process. To this end we require improved methods for producing excessively frequent alternations and enormous potentials. Shall we be able to obtain these by tapping the ether? If so, we may view the prospective exhaustion of our coal-fields with indifference; we shall at once solve the smoke question, and thus dissolve all possible coal rings.

Electricity seems destined to annex the whole field, not merely of optics, but probably also of thermotics.

Rays of light will not pass through a wall, nor, as we know only too well, through a dense fog. But electrical rays of a foot or two wave-length of which we have spoken will easily pierce such mediums, which for them will be transparent.

Another tempting field for research, scarcely yet attacked by pioneers, awaits exploration. I allude to the mutual action of electricity and life. No sound man of science indorses the assertion that "electricity is life"; nor can we even venture to speak of life as one of the varieties or manifestations of energy. Nevertheless, electricity has an important influence upon vital phenomena, and is in turn set in action by the living being – animal or vegetable. We have electric fishes – one of them the prototype of the torpedo of modern warfare. There is the electric slug which used to be met with in gardens and roads about Hornsey Rise; there is also an electric centiped. In the study of such facts and such relations the scientific electrician has before him an almost infinite field of inquiry.

The slower vibrations to which I have referred reveal the bewildering possibility of telegraphy without wires, posts, cables, or any of our present costly appliances. It is vain to attempt to picture the marvels of the future. Progress, as Dean Swift observed, may be too fast for endurance. Sufficient for this generation are the wonders thereof. – *Nature.*

Source Endnote

1 Speech delivered at the third annual dinner of the Institution of Electrical Engineers, London, November 13, 1891.

Editorial Headnote

11. *Social Evolution* (London: MacMillan and Co., 1894), pp. 1–7.

Benjamin Kidd's (1858–1916) publication of *Social Evolution* transformed the British sociologist into an internationally recognized author and thinker. Kidd circulated in Britain, Canada, South Africa, and the United States and argued for greater attention to be paid to the role of religion in the "evolution" of society. In addition to the *Social Evolution*, Kidd wrote histories and repudiations of western civilization. *Social Evolution*'s resonance typifies the evolution and social Darwinism publications of the late nineteenth century, popular in both Great Britain and the United States. He challenged the primacy of science in the public eye and argued that Christian altruism was the force that led Western Civilization to its current great state. For this volume, I selected pages from the first chapter of the book titled "The Outlook," in which he remarks on the history of the nineteenth century and the distressed state of an English society that has been incapable understanding itself through the explanations offered by people like Karl Marx (1818–1883) or scientific minds like Herbert Spencer (1820–1903). As we consider the imaginary that Victorians constructed around electricity, especially where the progress of civilization is concerned, I find it useful to note participants in that imaginary did not all affirm science as progress. Even still, they had to deal with electricity. Kidd, notes toward the end of this passage the marvellous and vast implications of electric communication and electric power on industry, the state, and human relationships. He equates electricity and the practical sciences with unlimited potential and expresses anxiety at what that might mean. For Kidd, and undoubtedly many others, electricity changed not only science but the fabric of society, a "nervous system" as he calls it. A system that he believes will replace old systems, construct new social classes, and organize the world with "bewildering intricacy."

11

SOCIAL EVOLUTION

Benjamin Kidd

Source: Benjamin Kidd, *Social Evolution* (London: MacMillan and Co., 1894), pp. 1–7.

CHAPTER I

THE OUTLOOK

To the thoughtful mind the outlook at the close of the nineteenth century is profoundly interesting. History can furnish no parallel to it. The problems which loom across the threshold of the new century surpass in magnitude any that civilisation has hitherto had to encounter. We seem to have reached a time in which there is abroad in men's minds an instinctive feeling that a definite stage in the evolution of Western civilisation is drawing to a close, and that we are entering on a new era. Yet one of the most curious features of the time is the almost complete absence of any clear indication from those who speak in the name of science and authority as to the direction in which the path of future progress lies. On every side in those departments of knowledge which deal with social affairs change, transition, and uncertainty are apparent. Despite the great advances which science has made during the past century in almost every other direction, there is, it must be confessed, no science of human society properly so called. What knowledge there is exists in a more or less chaotic state scattered under many heads; and it is not improbably true, however much we may hesitate to acknowledge it, that the generalisations which have recently tended most to foster a conception of the unity underlying the laws operating amid the complex social phenomena of our time, have not been those which have come from the orthodox scientific school. They have rather been those advanced by that school of social revolutionists of which Karl Marx is the most commanding figure. Judged by the utterances of her spokesmen, science, whose great triumph in the nineteenth century has been the tracing of the steps in the evolution of life up to human society, stands now dumb before the problems presented by society as it exists around us. As regards its further evolution she appears to have no clear message.

In England we have a most remarkable example of the attitude of science when she is appealed to for aid and enlightenment in those all-engrossing problems with which society is struggling. One of the monumental works of our time is the "Synthetic Philosophy" of Mr. Herbert Spencer begun early in the second half of

DOI: 10.4324/9781003295556-13

the century, and not yet completed. It is a stupendous attempt not only at the unification of knowledge, but at the explanation in terms of evolutionary science of the development which human society is undergoing, and towards the elucidation of which development it is rightly recognised that all the work of science in lower fields should be preliminary. Yet so little practical light has the author apparently succeeded in throwing on the nature of the social problems of our time, that his investigations and conclusions are, according as they are dealt with by one side or the other, held to lead up to the opinions of the two diametrically opposite camps of individualists and collectiviste into which society is slowly becoming organised.

From Mr. Herbert Spencer in England, who himself regards the socialistic tendencies of the times with dislike if not with alarm, and whose views are thus shared by some and opposed by others of his own followers, to Professor Schäffle in Germany, who regards the future as belonging to purified socialism, we have every possible and perplexing variety of opinion. The negative and helpless position of science is fairly exemplified in England by Professor Huxley, who in some of his recent writings has devoted himself to reducing the aims of the two conflicting parties of the day – individualists and socialists – to absurdity and impossibility respectively. These efforts are not, however, to be regarded as preliminary to an attempt to inspire us with any clear idea as to where our duty lies in the circumstances. After this onslaught his own faith in the future grows obscure, and he sends his readers on their way with, for guiding principle, no particular faith or hope in anything.[1]

Yet that the times are pregnant of great changes the least observant must be convinced. Even those who indulge in these destructive criticisms seem to be conscious of this. Professor Huxley himself, despite his negative conclusions, is almost as outspoken as a Nihilist in his dissatisfaction with the existing state of things. "Even the best of modern civilisations," said he recently, "appears to me to exhibit a condition of mankind which neither embodies any worthy ideal nor even possesses the merit of stability. I do not hesitate to express the opinion that if there is no hope of a large improvement of the condition of the greater part of the human family; if it is true that the increase of knowledge, the winning of a greater dominion over nature which is its consequence, and the wealth which follows upon that dominion are to make no difference in the extent and the intensity of want with its concomitant physical and moral degradation amongst the masses of the people, I should hail the advent of some kindly comet which would sweep the whole affair away as a desirable consummation."[2] It is the large body of thought which this kind of feeling inspires which is now stirring European society to its depths, and nothing is more certain than that it will have to be reckoned with. M. de Laveleye, a few years ago, put the feeling into words. The message of the eighteenth century to man was, he said, "Thou shalt cease to be the slave of nobles and despots who oppress thee; thou art free and sovereign." But the problem of our times is: "It is a grand thing to be free and sovereign, but how is it that the sovereign often starves? how is it that those who are held to be the source of power often cannot, even by hard work, provide themselves with the necessaries of life?"[3] Mr. Henry George only fairly presses the matter home by asking whither in such circumstances our progress is leading; for, "to educate men who must be condemned to poverty is but to make them restive; to base on a state of most glaring social inequality political institutions under which men are theoretically equal is to stand a pyramid on its apex."[4]

Those who wish to see the end of the present condition of society have, so far, taken most part in the argument. Those who have no desire for change are of the class which always waits for action rather than argument. But a large section of the community, probably the largest section, while remaining unconvinced by the arguments used and more or less distrusting the methods proposed, feel that some change is inevitable. It is with these will probably rest the decisive part in shaping the course of future events. But at present they simply sit still and wait. They have no indication as to the direction in which the right path lies. They look in vain to science and authority for any hint as to duty. They are without a faith, for there is at the present time no science of human society. Many of the spokesmen of science who concern themselves with social problems continue to speak and act as if they conceived that their duty to society was to take away its religious beliefs. But it is not that they have any faith of their own to offer instead; they apparently have themselves no grasp of the problems with which the world is struggling as best it can. Science has obviously herself no clear perception of the nature of the social evolution we are undergoing. She has made no serious attempt to explain the phenomenon of our Western civilisation. We are without any real knowledge of the laws of its life and development or of the principles which underlie the process of social evolution which is proceeding around us.[5]

To many the spirit of the French Revolution which caused so universal a feeling of unrest at the end of the last century seems to be again unloosed, and after an epoch of progress unexampled in the history of the world we would appear to have returned to the discussion of the ideals of society which moved men's minds at that period of upheaval. Nothing can, however, be more out of place than comparisons which are instituted between society one hundred years ago and at the present time. We have little in common with this past. It may be searched in vain for any clue to the solution of the problems which confront us in the future. The great political revolution which began one hundred years ago, and which has been in progress in England and on the Continent throughout the nineteenth century, has well-nigh attained its ends. The middle classes having succeeded in enfranchising themselves have been in turn driven to enfranchise the lower classes; and with the possession of universal education and universal suffrage, and the long list of measures tending the more fully to secure the political enfranchisement of the people which has accompanied them, this revolution is, to all intents, complete. We have in reality entered on a new stage of social evolution in which the minds of men are moving towards other goals; and those political parties which still stand confronting the people with remnants of the political programme of political equality are beginning to find that the world is rapidly moving beyond their standpoint.

In other directions, too, the changes have been vast. Since the beginning of the century applied science has transformed the world. Amongst the advanced nations, the great wave of industrial expansion which follows in its wake is slowly but inevitably submerging the old landmarks of society, and preparing for us a world where the old things, material and social as well as political, have passed away, and in which the experience of the past is no longer a reliable guide. The marvellous development of practical science, the revolution in industry which it has effected, the application of steam and electricity on an immense scale to machinery, the enormous extension of railways, telegraphs, and other means of rapid communication, the development of commerce to a degree never before imagined, are amongst the wonders of the present age. They are only the earnest apparently of the future. Even a superficial acquaintance with the means and methods of modern science can hardly fail to leave the conviction that no limit can be set to the possibilities of even the near future, and that the achievements of the past, extraordinary as they have been, are not improbably destined to be eclipsed at no distant date.

Source Endnotes

1 See his "Government: Anarchy or Regimentation," *Nineteenth Century,* May 1890. See also his *Social Diseases and Worse Remedies,* pp. 13–51.

2 "Government: Anarchy or Regimentation," by Professor Huxley, *Nineteenth Century,* May 1890.

3 "Communism:" by Emile de Laveleye, *Contemporary Review,* March 1890.

4 *Progress and Poverty,* Introduction.

5 So far the larger part of the most useful work of the century in the department of sociology appears to have been merely destructive. "It may be stated," said Mr. Leslie Stephen recently, "that there is no science of sociology properly scientific – merely a heap of vague empirical observations, too flimsy to be useful in strict logical inference." – *Presidential Address, Annual Meeting of the Social and Political Education League,* March 1892.

Editorial Headnote

12. "The Universality of Electricity," *Punch*, 1858, October 23, p. 165.

13. "Prometheus Unbound: Science in Olympus," *Punch Magazine*, December 9, 1878.

Since 1841, *Punch Magazine*'s satirical writing and cartoons provide an excellent insight into Victorian life and culture. Increasingly, the magazine has become an incredibly popular source for sociocultural historians studying nineteenth- and twentieth-century Great Britain. *Punch* launched on July 17, 1841, inspired by highly successful Parisian satirical weeklies like *Le Charivari.* The engraver Ebenezer Landells (1808–1860) believed a wood engraving satirical publication produced by the talent on London's Fleet Street to commend on the scandal, blackmail, and squabbles of the day could be highly successful. During the first years of the publication, the pages were mostly filled with radical political commentary and contained few illustrations. Increasingly, under illustrators like John Leech (1817–1864), the magazine became more well-known for its illustrations or cartoons. While the magazine began as means through which to critique the cultural and political establishment, during the magazine's heyday in the late nineteenth, they became the guardians of the establishment and Victorian moral high ground. One of the places in which this becomes most apparent is in the essayists' and illustrators' portrayal of electric science and technology. These productions provide us a way to appreciate the scope of the imagination and politics Victorians attached to electricity – ranging from marvel to extreme anxiety. The first is an essay titled "The Universality of Electricity," in which the illustrator George Cruikshank (1792–1878) produced a cartoon of a man recovering with glee from the application of electricity to tooth removal. The essay compares the possibilities of electricity in dentistry and the playing of pianos by electricity. The purpose of the essay is to ask, "What will not electricity do next?" On the one hand, perhaps electricity will do everything (except income taxes), but on the other, where does it stop? The same types of questions come in the next series of illustrations I selected from *Punch* almost 20 years later. In this volume, the frontispiece features Mr. Punch jumping on an arrangement of accumulators that look like they were modelled after Leyden Jars, with lids showing the astrological calendar, while also surrounded by electrical equipment. He is also holding an animal labeled *gymnotus*, referencing the torpedo fish or electrical eel that became famous in many nineteenth-century experiments. In the following pages, cartoons parody Thomas Edison's apocryphal inventions such as antigravitation underwear, allowing museum patrons to reach new heights in an art gallery, or the "telephonoscope," allowing parents in England to speak to their children in the Antipodes. These cartoons and the many others within other editions of *Punch* demonstrate that Victorians still understood electricity as a captivating phenomenon with the potential to reshape their experiences. At the same time, Mr. Punch illustrates how some electric technologies are superfluous and might weaken people because of the dependencies they create. The possibility of and anxiety around electricity ranged wildly in the press and readers.

12

THE UNIVERSALITY OF ELECTRICITY

Source: "The Universality of Electricity," *Punch*, 1858, October 23, p. 165.

OUR daily reading proves that Electricity is now fairly taking the circuit of the entire globe. No barrister goes so extensive a circuit, or talks so much with so little noise. The beauty of electricity is, that it talks without being heard, an accomplishment which ladies have not yet acquired.

Amongst the recent marvels of electricity, we have to record two, viz.: – pianoforte-playing, and tooth-drawing. We need not say that both performances are at times equally shocking. There are occasions when we would sooner have a tooth drawn, we think, than listen to an excruciating extraction of agonising sounds from the piano. So much depends upon the instrument, and the use that is made of it! According to your manipulation, it becomes either an instrument of pleasure, or an instrument of torture. But if it is occasionally horrible to hear *one* piano (and the horror occurs in every capital, that has any pretensions to be considered a PIANOPOLIS, more than once a day), consider what it must be to listen to five pianos *eodem tempore!* Electricity confers this terrible dexterity on every pianoforte practitioner. This quintuple achievement was actually accomplished by a MR. LÉON HUMAR, at the National Theatre at Brussels. The five pianos were lashed together with electric wires, which were connected with an electric battery in another room, and you had the whole lot of them jingling away at once. No necessity to encore a piece of music at this rate, when you take it in five times over at a single hearing. What a fearful power to vest in any man's hands! If electricity can do it with five pianos, why not with fifty? why not with five hundred? Where is it to stop? If that is all, we do not see what there is to prevent a pianist, who holds this electric accomplishment at his fingers'-ends, from performing in every capital of Europe at precisely the same time. Fancy LISTZ going through his pianofortefistic gymnastics on five hundred Broadwoods, and being heard simultaneously, without the interval of scarce a vibration, all over the world! We should not be safe anywhere. He might pursue us into the very centre of the Desert. What exquisite revenge he might take on his detractors! He might maliciously place a piano on each side even of *us,* put one over our heads, and another under our feet, and by playing upon them, unseen by us, and unknown to us, some 2000 miles off, send us raving mad in less than half-an-hour! We hope electricity will never lend itself to such base machinations.

DOI: 10.4324/9781003295556-14

We have heard of men riding on five horses; but we do not think that a pianoforte-player should try to emulate a horse-rider by playing upon five pianos all at once. One piano at a time is quite enough, and frequently too much; but bring the electric battery to play upon an allied army of Collards, and we will not answer for the world being strong enough to stand the shock. Further, we have a presentiment that it would be shivered to little bits by sheer force of melody, as we have seen a magnificent barley-sugar temple on a supper-table totter and fall with a crackling crash, into a chaos of golden ruins, by the violence of the stamping music overhead. So it would be with the world! After one five-hundred pianoforte *séance,* there would be nothing left of it but an immense dust-heap, on the top of which men and women would be lying like so many oyster-shells and lobster's claws.

As for the toothdrawing by electricity, we do not jump to it so easily. It is actually done, however, in far less time than you can think as to what you shall have for to-day's dinner? A single tooth, we suppose, is charged at the rate of a single message; a double tooth doubtlessly counts for two. We wonder if the same shock that pulls out the tooth can convey a message to tell your wife that it is OUT! and "out (as GEORGE CRUIKSHANK's drawing informs us with jumping glee) in less than a minute." The description of this new style of dentition neglects to inform us, whether a new set of teeth can be supplied by the same process.

The anticipated benefit of the one discovery tends greatly to balance the dreaded evil of the other. If pianoforte playing by electricity threatens to increase largely human suffering, at all events dental surgery by the same invisible agency promises to considerably alleviate it. It is all a question of nerves, and in the electric contest between ears and teeth, let us hope the molars will have it.

The question is, what will not electricity do next? We do not despair of the good time coming (and it has been a long time on the road), when we shall be able to sit quietly in our arm-chair and electricity will do everything for us. It will cook our dinner, sew on our buttons, write our letters, make our clothes, whip our children, black our boots, shave our stubbly chins, and even help us to a pinch of snuff, if we only wish it. We almost believe it will in time so far reach mortal perfection as to carry us up to bed, undress us, tuck us up, and blow out the candle, when we are too tired, or indifferent, to do it ourselves. But there is one thing, we are afraid, that it never will do, and that is, help us to pay our Income-Tax.

Alas! there is a limit even to electricity!

13

PROMETHEUS UNBOUND

Science in Olympus (illustration)

Source: "Prometheus Unbound: Science in Olympus," *Punch Magazine*, December 9, 1878.

Editorial Headnote

14. "Telectroscopy," *The Electrical Engineer*, March 25, 1898, p. 354.

The possibilities ascribed to Edison's telephonoscope did not only feature in the pages of *Punch*, but also found space in *The Electrical Engineer*, a weekly British periodical featuring electrical engineering news, notes, and papers founded in 1885, not to be confused with the American publication of the same title which was founded as the *World's Fair & Electrical Engineering* in concert with the 1893 World's Columbian Exhibition. In particular late nineteenth-century note, the publication features two insightful pieces on the future electric technologies: telectroscopy and space telegraphy. In the first, the authors describe the work of Herr Jan Szczepanik (1872–1926), a Polish inventor, known for his work on fabric and photography. The work telectroscope, or sometimes electroscope, is often anachronistically discussed as an early television and associated with inventors and scientists like Szczepanik, Alexander Graham Bell (1847–1922), George Carey (1851–1906), and Constantin Senlecq (1842–1934). Often, what they aimed to produce, actually produced, or patented did not aim to create video communication technology. And more importantly, their invention and existence remained contentious. For example, many people doubted whether Szczpanik's device even existed, which you can pick up in the note. However, the way the author and Szczpanik imagined the telectroscope typifies the seemingly limitless potential people assigned to electrical technology at the time. Szczpanik's device, according to the Vienna correspondent mentioned here, held numerous possibilities but also Not. This is a subtle but important shift away from the imagined possibilities of electrical technologies in some of the pieces mentioned earlier in the volume that allowed a great deal of ambiguity. Szczpanik needed the device to replace the telegraph and graft on to people's existing expectations of electric power. Following this piece, the author reports on a lecture given by Oliver Lodge (1851–1940) to the Engineering Society at Liverpool on the transmission of electric waves by electric oscillation. Lodge noted that the mechanisms behind these waves had been explored by Joseph Henry (1797–1878), Alexander von Humboldt (1769–1859), Lord Kelvin (1824–1907), James Clerk Maxwell (1831–1879), George Fitzgerald (1851–1901), and Heinrich Hertz (1857–1894) for decades and might indeed lead to a system of wireless telegraphy covering vast distance without burdensome wires. And indeed, three years later, Guglielmo Marconi (1874–1937) demonstrated such a system from the British Post Office. Again, the article showcases a hope for the electric technologies of the future, but these imaginaries, especially the wireless telegraph, needed to be grounded within a celebrated long-running scientific endeavor, and, like the telectroscope, be useful.

14

TELECTROSCOPY

Source: "Telectroscopy," *The Electrical Engineer*, March 25, 1898, p. 354.

The *Telegraph* is still cocksure as to the value of the discovery of Herr Szczepanik, in that it gives the following from its Vienna correspondent, and among other headlines says, "Further details of the telectroscope." We fail to find the details, but perhaps our readers can do so: "The inventor of the telectroscope, Herr Szczepanik, will introduce his discovery in the course of the next few days to a select circle of scientific men and journalists. In contradicting certain remarks in the Press, according to which the telectroscope has been described by experts as a discovery which at present has no practical utility, Herr Szczepanik declared to-day that the telectroscope possesses far greater practical importance and possibilities of usefulness than is supposed. The apparatus, says Herr Szczepanik, not only can produce pictures from a long distance, but will make the entire system of telegraphy as it now exists superfluous. Then, as well as portraying real proceedings in the shape of pictures, the telectroscope will also show copies of manuscripts and prints at the remotest distances in a moment of time, and fix at one stroke photographic facsimiles of the same on a sensitive plate or sensitive paper. This transmission will extend, as regards distance, as far as will the telephone. Respecting this telegraphy of the future, Herr Szczepanik declares that it is possible, for example, to place a copy of a newspaper in the apparatus which has been prepared by him and is now ready for use, and within a few seconds, at such a distance, say, as from Vienna to Berlin, a photographic facsimile of this newspaper would be yielded."

DOI: 10.4324/9781003295556-16

Part 3

EXPERIMENTS

Editorial Headnote

15. "On the Electricity Excited by the Mere Contact of Conducting Substances of Different Kinds," *The Philosophical Magazine*, September 1800, pp. 289–311.

As natural philosophy, or science, increasingly became institutionalized, one of the ways members of organizations like the Royal Society or Académie des sciences communicated was through correspondence. In addition to treatises, experiments, or proofs, these letters functioned as a mechanism through which rapidly shifting theories, such as those concerning electrical phenomena, could be explored and vetted by the authors and readers of various scientific communities. Often, letters were featured in multiple publications. For instance, this letter shown here in English was translated from an earlier publication in French. Alessandro Volta (1745–1827), a professor of natural philosophy at the University of Pavia, wrote this letter to Joesph Banks (1743–1820), who at the time served as the president of the Royal Society. Prior to this letter, Volta, along with many other natural philosophers, built his academic career around the study of electric phenomena and what we would today call electrochemistry. He was perhaps most famous for his work on the voltaic pile, or battery. This letter, which Banks would have read aloud to the Royal Society, is where Volta describes the voltaic apparatus capable of "generating" an electric current. After remarking on the striking and astonishing facets of his experiments, he argues that metal, through mere contact with other material, can generate a current. He compares what such a device might do like the Leyden jar (or flask). He then lays out the making of the device by stacking plates of copper and zinc with pasteboard (cardboard). The pile is then capable of producing a shock, but unlike a Leyden jar or other such device it has no need of being charged by "foreign electricity." He then proceeds to give an incredibly detailed description of its construction and the various effects it has on the body. Apart from solidifying and exciting natural philosophers about the potential of batteries, Volta's letter highlights a major shift in how people conceptualized electricity. Volta diverged from Galvani and Humbolt by suggesting that a distinct form of electricity intrinsic to animals existed. He used his pile to show that the same, or true electricity, could be created with inanimate materials. Although debates about Galvanism and animal electricity raged, Volta's letter and many of his subsequent publications mark a shift in how people conducted electrical experiments.

15

ON THE ELECTRICITY EXCITED BY THE MERE CONTACT OF CONDUCTING SUBSTANCES OF DIFFERENT KINDS[1]

Alexander Volta to Joseph Banks

Source: Alexander Volta to Joseph Banks, "On the Electricity Excited by the Mere Contact of Conducting Substances of Different Kinds," *The Philosophical Magazine*, September 1800, pp. 289–311.

Como in the Milanese, March 20, 1800.

AFTER a long silence, for which I shall offer no apology, I have the pleasure of communicating to you, and through you to the Royal Society, some striking results I have obtained in pursuing my experiments on electricity excited by the mere mutual contact of different kinds of metal, and even by that of other conductors, also different from each other, either liquid or containing some liquid, to which they are properly indebted for their conducting power. The principal of these results, which comprehends nearly all the rest, is the construction of an apparatus having a resemblance in its effects (that is to say, in the shock it is capable of making the arms, &c. experience) to the Leyden flask, or, rather, to an electric battery weakly charged acting incessantly, which should charge itself after each explosion; and, in a word, which should have an inexhaustible charge, a perpetual action or impulse on the electric fluid; but which differs from it essentially both by this continual action, which is peculiar to it, and because, instead of consisting, like the common electric jars and batteries, of one or more insulating plates or thin strata of those bodies which are alone thought to be *electric,* armed with conductors, or bodies called *non-electric,* this new apparatus is formed merely of several of the latter bodies, chosen from among those which are the best conductors, and therefore the most remote, as has hitherto been believed, from the electric nature. The apparatus to which I allude, and which will, no doubt, astonish you, is only the assemblage of a number of good conductors of different kinds arranged in a certain manner. Thirty, forty, sixty, or more pieces of copper, or rather silver, applied each to a piece of tin, or zinc, which is much better, and as many strata of water, or any other liquid which may be a better conductor, such as salt water, ley, &c. or pieces of pasteboard, skin, &c. well soaked in these liquids; such strata

DOI: 10.4324/9781003295556-18

interposed between every pair or combination of two different metals in an alternate series, and always in the same order of these three kinds of conductors, are all that is necessary for constituting my new instrument, which, as I have said, imitates the effects of the Leyden flask, or of electric batteries, by communicating the same shock as these do; but which, indeed, is far inferior to the activity of these batteries when highly charged, either in regard to the force and noise of the explosions, the spark, the distance at which the discharge may be effected, &c. as it equals only the effects of a battery very weakly charged, though of immense capacity: in other respects, however, it far surpasses the virtue and power of these batteries, as it has no need, like these, of being previously charged by means of foreign electricity, and as it is capable of giving a shock every time it is properly touched, however often it may be.

To this apparatus, much more similar at bottom, as I shall show, and even such as I have constructed it, in its form to the *natural electric organ* of the torpedo or electric eel, &c. than to the Leyden flask and electric batteries, I would wish to give the name of the *artificial electric organ:* and, indeed, is it not, like it, composed entirely of conducting bodies? Is it not also active of itself without any previous charge, without the aid of any electricity excited by any of the means hitherto known? Does it not act incessantly, and without intermission? And, in the last place, is it not capable of giving every moment shocks of greater or less strength, according to circumstances – shocks which are renewed by each new touch, and which, when thus repeated or continued for a certain time, produce the same torpor in the limbs as is occasioned by the torpedo, &c.?

I shall now give a more particular description of this apparatus and of others analogous to it, as well as of the most remarkable experiments made with them.

I provide a few dozens of small round plates or disks of copper, brass, or rather silver, an inch in diameter more or less (pieces of coin for example), and an equal number of plates of tin, or, what is better, of zinc, nearly of the same size and figure. I make use of the term *nearly,* because great precision is not necessary, and the size in general, as well as the figure of the metallic pieces, is merely arbitrary: care only must be taken that they may be capable of being conveniently arranged one above the other, in the form of a column. I prepare also a pretty large number of circular pieces of pasteboard, or any other spongy matter capable of imbibing and retaining a great deal of water or moisture, with which they must be well impregnated in order to ensure success to the experiments. These circular pieces of pasteboard, which I shall call moistened disks, I make a little smaller than the plates of metal, in order that, when interposed between them, as I shall hereafter describe, they may not project beyond them.

Having all these pieces ready in a good state, that is to say, the metallic disks very clean and dry, and the non-metallic ones well moistened with common water, or, what is much better, falt water, and slightly wiped that the moisture may not drop off, I have nothing to do but to arrange them, a matter exceedingly simple and easy.

I place then horizontally, on a table or any other stand, one of the metallic pieces, for example one of silver, and over the first I adapt one of zinc; on the second I place one of the moistened disks, then another plate of silver followed immediately by another of zinc, over which I place another of the moistened disks. In this manner I continue coupling a plate of silver with one of zinc, and always in the same order, that is to say, the silver below and the zinc above it, or *vice versa,* according as I have begun, and interpose between each of these couples a moistened disk. I continue to form, of several of these stories, a column as high as possible without any danger of its falling.

But, if it contain about twenty of these stories or couples of metal, it will be capable not only of emitting signs of electricity by Cavallo's electrometer, assisted by a condenser, beyond ten or fifteen degrees, and of charging this condenser by mere contact so as to make it emit a spark, &c. but of giving to the fingers with which its extremities (the bottom and top of the column) have been touched several small shocks, more or less frequent, according as the touching has been repeated. Each of these shocks has a perfect resemblance to that slight shock experienced from a Leyden flask weakly charged, or a battery still more weakly charged, or a torpedo in an exceedingly languishing state, which imitates still better the effects of my apparatus by the series of repeated shocks which it can continually communicate.

To obtain such slight shocks from this apparatus which I have described, and which is still too small for great effects, it is necessary that the fingers, with which the two extremities are to be touched at the same time, should be dipped in water, so that the skin, which otherwise is not a good conductor, may be well moistened. To succeed with more certainty, and receive stronger shocks, a communication must be made, by means of a metallic plate sufficiently large, or a large metallic wire, between the bottom of the column (that is to say, the lower piece of metal,) and water contained in a bason or large cup, in which one, two, or three fingers, or the whole hand is to be immersed, while you touch the top or upper extremity (the uppermost or one of the uppermost plates of the column) with the clean extremity of another metallic plate held in the other hand, which must be very moist, and embrace a large surface of the plate held very fast. By proceeding in this manner, I can obtain a small pricking or slight shock in one or two articulations of a finger immersed in the water of the bason, by touching, with the plate grasped in the other hand, the fourth or even third pair of metallic pieces. By touching then the fifth, the sixth, and the rest in succession till I come to the last, which forms the head of the column, it is curious to observe how the shocks gradually increase in force. But this force is such, that I receive from a column formed of twenty pairs of pieces (not more) shocks which affect the whole finger with considerable pain if it be immersed alone in the water of the bason; which extend (without pain) as far as the wrist, and even to the elbow, if the whole hand, or the greater part of it, be immersed; and are felt also in the wrist of the other hand.

I still suppose that all the necessary attention has been employed in the construction of the column, and that each pair or couple of metallic pieces, resulting from a plate of silver applied over one of zinc, is in communication with the following couple by a sufficient stratum of moisture, consisting of salt water rather than common water, or by a piece of pasteboard, skin, or any thing of the same kind well impregnated with this salt water. The disk must not be too small, and its surface must adhere closely to those of the metallic plates between which it is placed. This exact and extensive application of moistened disks is very important, whereas the metallic plates of each pair may only touch each other in a few points, provided that their contact is immediate.

All this shows that, if the contact of the metals with each other in some points only be sufficient (as they are excellent conductors) to give a free passage to a moderately strong current of electricity, the case is not the same with liquids, or bodies impregnated with moisture, which are conductors much less perfect; and which, consequently, have need of more ample contact with metallic conductors, and still more with each other, in order that the electric fluid may easily pass, and that it may not be too much retarded in its course; especially when it is moved with very little force, as in the present case.

In a word, the effects of my apparatus, that is to say, the shocks felt, are considerably more sensible in proportion as the temperature of the ambient air, or that of the water or moistened disks which enter into the composition of the column, and that of the water even in the bason, is warmer, as heat renders the water a better conductor. But almost all the salts, and particularly common salt, will render it a still better. This is one of the reasons, if not the only one, why it is so advantageous that the water of the bason, and, above all, that interposed between each pair of metallic plates, as well as the water with which the circular pieces of pasteboard are impregnated, &c. should be salt water, as already observed.

But all these means and all these attentions have only a limited advantage, and will never occasion your receiving very strong shocks as long as the apparatus consists but of one column, formed only of twenty pair of plates, even though they may consist of the two metals properest for these experiments, *viz.* silver and zinc; for if they were silver and lead, or tin, or copper and tin, the half of the effect would not be produced, unless the weaker effect of each pair were supplied by a much greater number. What really increases the electric power of this apparatus, and to such a degree as to make it equal or surpass that of the torpedo or electric eel, is the number of plates arranged in such a manner, and with the attention before mentioned. If to the twenty pairs above described twenty or thirty others be added disposed in the same order, the shocks which may be communicated by a column lengthened in this manner will be much stronger, and extend to both arms as far as the shoulder; and especially of that, the hand of which has been immersed in the water: this hand, with the whole arm, will remain more or less benumbed, if by frequently renewing the touches these shocks be made to succeed each other rapidly, and without intermission. This will be the case if the whole hand, or the greater part of it, be immersed in the water of the bason; but if only one finger be immersed, either wholly or in part, the shocks being almost entirely concentrated in it alone, will become so much the more painful, and so acute as to be scarcely supportable.

It may readily be conceived that this column, formed of forty or fifty couples of metals, which gives shocks more than moderate to both the arms of one person, is capable of giving sensible shocks also to several persons, holding each other by the hands (sufficiently moist) so as to form an uninterrupted chain.

I shall now return to the mechanical construction of my apparatus, which is susceptible of several variations, and describe not all those which I have invented or made, either on a small or a large scale, but only a few, which are either curious or useful, which exhibit some real advantage, as being easier or sooner constructed, and which are certain in their effects, or can be longer preserved in good order.

I shall begin by one which, uniting nearly all these advantages, differs most in its figure from the columnar apparatus above described, but which is attended with the inconvenience of being much more voluminous. This new apparatus, which I shall call a *couronne de tasses* (a chain of cups), is represented Plate VIII. fig. 1.

I dispose, therefore, a row of several basons or cups of any matter whatever, except metal, such as wood, shell, earth, or rather glass (small tumblers or drinking glasses are the most convenient), half filled with pure water, or rather salt water or ley: they are made all to communicate by forming them into a sort of chain, by means of so many metallic arcs, one arm of which, S*a,* or only the extremity S, immersed in one of the tumblers, is of copper or brass, or rather of copper plated with silver; and the other, *Za,* immersed into the next tumbler, is of tin, or rather of zinc. I shall here observe, that ley and other alkaline liquors are preferable when one of the metals to be immersed is tin: salt water is preferable when it is zinc. The two metals of which each arc is composed, are soldered together in any part above that which is immersed in the liquor, and which must touch it with a surface sufficiently large: it is necessary therefore that this part should be a plate of an inch square, or very little less; the rest of the arc may be as much narrower as you choose, and even a simple metallic wire. It may also consist of a third metal different from the two immersed into the tumblers, since the action on the electric fluid which results from all the contacts of several metals that immediately succeed each other, or the force with which this fluid is at last impelled, is absolutely the same, or nearly so, as that which it would have received by the immediate contact of the first metal with the last without any intermediate metals, as I have ascertained by direct experiments, of which I shall have occasion to speak hereafter.

A series of 30, 40, or 60 of these tumblers connected with each other in this manner, and ranged either in a straight or curved line, or bent in every manner possible, forms the whole of this new apparatus, which at bottom and in substance is the same as the other columnar one above described; as the essential part, which consists in the immediate communication of the different metals which form each couple, and the mediate communication of one couple with the other, *viz.* by the intervention of a humid conductor, exist in the one as well as the other.

In regard to the manner of trying these tumblers, and the different experiments for which they may be employed, there is no need of saying a great deal after the ample explanation I have already given respecting the columnar apparatus. It may be readily comprehended, that to obtain a shock it will be sufficient to immerse one hand into one of the tumblers, and a singer of the other hand into another of the tumblers at a considerable distance from the former: that this shock will be stronger the further these glasses are from each other; that is to say, in proportion to the number of the intermediate glasses, and consequently, that the strongest shock will be received when you touch the first and last end of the chain. It will be readily comprehended also, how and why the experiments will succeed much better by grasping and holding fast in one hand, well moistened, a pretty large plate of metal (in order that the communication may be more perfect, and formed in a great number of points), and touching with this plate the water in the tumbler, or rather the metallic arc, while the other is immersed in the other distant tumbler, or touches with a plate, grasped in the like manner, the arc of the latter. In a word, one may comprehend and even foresee the success of a great variety of experiments which may be made with this apparatus or chain of cups much more easily, and in a manner more evident, and which, if I may be allowed the expression, speak more to the eyes than those with the columnar apparatus. I shall therefore forbear from describing a great number of these experiments, which may be easily guessed, and shall relate only a few which are no less instructive than amusing.

Let three twenties of these tumblers be ranged, and connected with each other by metallic arcs, but in such a manner, that, for the first twenty, these arcs shall be turned in the same direction; for example, the arm of silver turned to the lest, and the arm of zinc to the right; and for the second twenty in a contrary direction, that is to say, the zinc to the left, and the silver to the right: in the last place, for the third twenty, the silver to the left, as is the case in regard to the first. When every thing is thus arranged, immerse one finger in the water of the first tumbler, and, with the plate grasped in the other hand, as above directed, touch the first metallic arc (that which joins the first tumbler to the second), then the other arc which joins the second and third tumbler, and so on, in succession, till you have touched them all. If the water be very salt and luke-warm, and the skin of the hands well moistened and softened, you will already begin to feel a slight shock in the finger when you have touched the fourth or fifth arc (I have experienced it sometimes very distinctly by touching the third), and by successively proceeding to the sixth and the seventh, &c. the shocks will gradually increase in force to the twentieth arc, that is to say, to the last of those turned in the same direction; but by proceeding onwards to the 21st, 22d, 23d, or 1st, 2d, 3d, of the second twenty, in which they are all turned in a contrary direction, the shocks will each time become weaker, so that at the 36th or 37th, they will be imperceptible, and be entirely null at the 40th, beyond which (and beginning the third twenty, opposed to the second and analogous to the first,) the shocks will be imperceptible to the 44th or 45th arc; but they will begin to become sensible, and to increase gradually, in proportion as you advance to the 60th, where they will have attained the same force as that of the 20th arc.

If the twenty arcs in the middle were all turned in the same direction as the preceding twenty and the following twenty, that is to say, if the whole 60 conspired to impel the electric fluid in the same direction, it may readily be comprehended how much greater the effect will be at the end, and how much stronger the shock; and it may be comprehended, in general, to what point it must be weakened in all cases where a greater or smaller number of these forces act contrary to each other by an inverted position of metals.

If the chain be in any part interrupted, either by one of the tumblers being empty of water, or one of the metallic arcs being removed or divided into two pieces, you will receive no shock when you immerse your finger into the water of the first and another into that of the last vessel; but you will have it strong or weak, according to circumstances (leaving these fingers immersed), at the moment when the interrupted communication is restored; at the moment when another person shall immerse into the two tumblers, where the arc is wanting, two of his fingers (which will also receive a slight shock), or rather, when he shall immerse the same arc which has been taken away, or any other; and in the case of the arc separated into two pieces, at the moment when these pieces are again brought into mutual contact (in which case the shock will be stronger than in any other); and, lastly, in the case of the empty tumbler, at the moment when water poured into it shall rise to the two metallic arms immersed in this cup which before were dry.

When the chain of cups is of sufficient length, and capable of giving a strong shock, you will experience one, though much weaker, even though you keep immersed two fingers, or the two hands, in one bason of water of pretty large size, in which the first and last metallic arcs are made to terminate: provided that either of these hands thus immersed, or rather both of them, be kept respectively in contact, or nearly in contact, with these arcs, you will, I say, experience a shock at the momen when (the chain being interrupted in any part) the communication is restored, and the circle completed in any of the ways before mentioned. One might be surprised that in this circle the electric current having a free passage through an uninterrupted mass of water, that which fills the bason, should quit this good conductor to throw itself and pursue its course through the body of the person who holds his hands immersed in the same water, and thus to take a longer passage. But the surprise will cease if we reflect, that living and warm animal substances, and above all, their humours, are, in general, better conductors than water. As the body, then, of the person who immerses his hands in the water, affords an easier passage than this water does to the electric current, the latter must prefer it though a little longer. In a word, the electric fluid, when it must traverse imperfect conductors in a large quantity, and particularly moist conductors, has a propensity to extend itself in a larger stream, or to divide itself into several, and even to pursue a winding course, as it thereby finds less resistance than by following one single channel, though shorter; in the present case it is only a part of the electric current, which, leaving the water, pursues this new route through the body of the person, and traverses it from the one arm to the other: a greater or less part passes through the water in the vessel. This is the reason why the shock experienced is much weaker than when the electric current is not divided when the person alone forms the communication between one arc and another, &c.

From these experiments one might believe, that when the torpedo wishes to communicate a shock to the arms of a man or to animals which touch it, or which approach its body under the water (which shock is much weaker than what the fish can give out of the water), it has nothing to do but to bring together some of the parts of its electric organ in that place, where, by some interval, the communication is interrupted, to remove the interruptions from between the columns of which the said organ is formed, or from between its membranes in the form of thin disks, which lie one above the other from the bottom to the summit of each column: it has, I say, nothing to do but to remove these interruptions in one or more places, and to produce there the requisite contact, either by compressing these columns, or by making some moisture to flow in between the pellicles or diaphragms which have been separated, &c. This is what may be, and what I really conclude to be, the task of the torpedo when it gives a shock; for all the rest, the impulse and movement communicated to the electric fluid, is only a necessary effect of its singular organ, formed, as is seen, of a very numerous series of conductors, which I have every reason to believe sufficiently different from each other to be *exciters* of the electric fluid by their mutual contacts; and to suppose them ranged in a manner proper for impelling that fluid with a sufficient force from top to bottom, or from the bottom to the top, and for determining a current capable of producing the shock, &c. as soon and as often as all the necessary contacts and communications take place.

But let us now leave the torpedo, and its *natural electric organ,* and return to the *artificial electric organ* of my invention, and particularly to my first *columnar apparatus,* that which imitates the first even in its form (for that composed of tumblers is different in that respect). I might say something also in regard to the construction of the said apparatus with tumblers or *a chain of glasses*; for example, that the first and last tumbler should be of such a size that, when necessary, the whole hand might be immersed in it, &c.; but, to enter into all these details, would require too much time.

In regard to the columnar apparatus, I endeavoured to discover the means of lengthening it a great deal by multiplying the metallic plates in such a manner as not to tumble down; and I discovered, besides others, the following, which are represented in the annexed figures. (Plate VIII. fig. 2, 3, 4.)

In Fig. 2, *m m m m* are rods, three, four, or more in number, which rise from the bottom of the column, and confine, as in a cage, the plates or disks, placed each above the other in such number and to such a height as you choose, and which thus prevent them from falling. The rods may be of glass, wood, or metal, only that, in the last case, you must prevent them from coming into immediate contact with the plates; which may be done either by covering each of them with a glass tube, or interposing between them and the column a few stripes of wax cloth, oiled paper, or even plain paper, and, in a word, any other body that may either be a *cohibent* or a bad conductor: wood or paper will be sufficiently so for our purpose, provided only that they are not very damp or moist.

But the best expedient, when you wish to form an apparatus to consist of a great number of plates, above 60, 80, or 100 for example, is, to divide the column into two or more, as seen Fig. 3 and 4, (Plate VIII.) where the pieces all have their respective positions and communication as if there were only one column. Fig. 4, as well as Fig. 3, may indeed be considered as a bent column.

In all these figures the different metallic plates are denoted by the letters S and Z (which are the initials of silver and zinc); and the *moistened disks* (of pasteboard, skin, &c. interposed between each pair of metals), are represented by a black stratum. The plates of metal may either be laid simply upon each other and so brought into union in an indefinite number of points, or they may be soldered together. It is altogether indifferent whichever of these methods be followed. *cc, cc, cc,* are the metallic plates which form a communication between each column, or section of a column, and another; and *bb, bb, bb,* are the basons of water in communication with the lower part or extremities of these columns.

An apparatus thus prepared is exceedingly convenient without being bulky; and it might be rendered portable, with still more ease and safety, by means of circular cases or tubes, in which each column might be inclosed and preserved. It is only to be regretted that it does not long continue in a good state: the moistened disks become dry in one or two days to such a degree that they must be again moistened; which, however, may be done without taking to pieces the whole apparatus, by immersing the columns, completely formed, in water, and wiping them, when taken out some time after, with a cloth, or in any other manner.

The best method of making an instrument as durable as can be wished, would be, to inclose and confine the water interposed between each pair of metals, and to fix these metallic plates in their places by enveloping the whole column with wax or pitch: but this would be somewhat difficult in the execution, and would require a great deal of patience. I have, however, succeeded; and have formed in this manner two cylinders consisting of twenty pair of metals, which can still be employed though made several weeks, and which, I hope, will be serviceable for months.

These cylinders are attended with this advantage, that they may be employed for experiments either in an erect, inclined, or lying position, according as you choose, or even immersed in water, provided the top of it be above the surface of the fluid: they might also give a shock when entirely immersed if they contained a greater number of plates, or if several of these cylinders were joined together, and if there were any interruption that could be removed at pleasure, &c. by which means these cylinders would have a pretty good resemblance to the electric eel; and, to have a better resemblance to it even externally, they might be joined together by pliable metallic wires or screw-springs, and then covered with a skin terminated by a head and tail properly formed, &c.

The effects sensible to our organs produced by an apparatus formed of 40 or 50 pair of plates (and even by a smaller, if one of the metals be silver or copper and the other zinc,) are reduced merely to shocks: the current of the electric fluid, impelled and excited by such a number and variety of different conductors, silver, zinc, and water, disposed alternately in the manner above described, excites not only contractions and spasms in the muscles, convulsions more or less violent in the limbs through which it passes in its course; but it irritates also the organs of taste, sight, hearing, and feeling, properly so called, and produces in them sensations peculiar to each.

And first, in regard to the sense of feeling: If, by means of an ample contact of the hand (well moistened) with a plate of metal, or rather, by immersing the hand to a considerable depth in the water of the bason, I establish on one side a good communication with one of the extremities of my *electro-motive* apparatus, (we must give new names to instruments that are new not only in their form, but in their effects or the principle on which they depend); and on the other I apply the forehead, eye-lid, tip of the nose, also well moistened, or any other part of the body where the skin is very delicate: if I apply, I say, with a little pressure, any one of these delicate parts, well moistened, to the point of a metallic wire, communicating properly with the other extremity of the said apparatus, I experience, at the moment that the conducting circle is completed, at the place of the skin touched, and a little beyond it, a blow and a prick, which suddenly passes, and is repeated as many times as the circle is interrupted and restored; so that, if these alternations be frequent, they occasion a very disagreeable quivering and pricking. But if all these communications continue without these alternations, without the least interruption of the circle, I feel nothing for some moments; afterwards, however, there begins at the part applied to the end of the wire, another sensation, which is a sharp pain (without shock), limited precisely by the points of contact, a quivering, not only continued, but which always goes on increasing to such a degree, that in a little time it becomes insupportable, and does not cease till the circle is interrupted.

What proof more evident of the continuation of the electric current as long as the communication of the conductors forming the circle is continued? – and that such a current is only suspended by interrupting that communication? This endless circulation of the electric fluid (this *perpetual motion*) may appear paradoxical and even inexplicable, but it is no less true and real; and you feel it, as I may say, with your hands. Another evident proof may be drawn from this circumstance, that in such experiments you often experience, at the moment when the circle is suddenly interrupted, a shock, a pricking, an agitation, according to circumstances, in the same manner as at the moment when it is completed; with this only difference, that these sensations, occasioned by a kind of reflux of the electric fluid, or by the shock which arises from the sudden suspension of its current, are of less strength. But I have no need, and this is not the place to bring forward proofs of such an endless circulation of the electric fluid in a circle of conductors, where there are some, which, by being of a different kind, perform, by their mutual contact, the office of exciters or *movers:* this proposition, which I advanced in my first researches and discoveries on the subject of galvanism, and always maintained by supporting them with new facts and experiments, will, I hope, meet with no opposers.

Recurring to the sensation of pain which is felt in the experiments above described, I must add, that if this pain be very strong and pricking in the parts covered by the skin, it is much more so in those where the skin has been taken off – in recent wounds for example. If by chance there should be a small incision or bit of the skin rubbed off in the singer which I immerse in the water that communicates with one of the extremities of the *electro-motive* apparatus, I experience there a pain so acute, when, by establishing the proper communication with the other extremity, I complete the circle, that I must soon desist from the experiment; that is to say, must withdraw my finger, or interrupt the circle in some other manner. I will say more; that I cannot even endure it above a few seconds when the part of the apparatus which I put in play, or the whole apparatus, contains only twenty pair of plates, or about that number.

One thing, which I must still remark, is, that all these sensations of pricking and pain are stronger and sharper, every thing else being equal, when the part of the body which is to feel them is towards the negative electricity; that is to say, placed in such a manner in the conducting circle, that the electric fluid traversing that circle is not directed towards that sensible part, does not advance towards it, and enter from the outside inwards, but takes its direction from the inside outwards; in a word, that it issues from it: in regard to which it is necessary to know, of the two metals that enter by pairs into the construction of the machine, which is the one that gives off to the other. But I had already determined this respecting all the metals by other experiments, published a long time ago at the end of my first memoirs on galvanism. I shall therefore say nothing further here, than that the whole is completely confirmed by the experiments, equally and still more demonstrative and striking, with which I am at present employed.

In regard to the sense of taste, I had before discovered, and published in these first memoirs, where I found myself obliged to combat the pretended animal electricity of Galvani, and to declare it an external electricity moved by the mutual contact of metals of different kinds, – I had discovered, I say, in consequence of this power which I ascribed to metals, that two pieces of these different metals, and particularly one of silver and one of zinc, applied in a proper manner, excited at the tip of the tongue very sensible sensations of taste; that the taste was decidedly acid, if, the tip of the tongue being turned towards the zinc, the electric current proceeded against it, and entered it; and that another taste, less strong but more disagreeable, acrid, and inclining to alkaline, was felt, if (the position of the metals being reversed) the electric current issued from the tip of the tongue; that these sensations continued and received even an increase for several seconds, if the mutual contact of the two metals was maintained, and if the conducting circle was nowhere interrupted. But when I have said here, that exactly the same phenomena take place when you try, instead of one pair of these metallic pieces, an assemblage of several of them ranged in the proper manner; and that the said sensations of taste, whether acid or alkaline, increase but a little with the number of these pairs, I have said the whole. It only remains for me to add that, if the apparatus put in play for these experiments on the tongue be formed of a sufficiently large number of metallic pairs of this kind, for example, if it contain 30, 40, or more, the tongue experiences not only the sensation of taste already mentioned, but, besides that, a blow which it receives at the moment when the circle is completed, and which occasions in it a pricking more or less painful, but fleeting, followed some moments after by a durable sensation of taste. This blow produces even a convulsion or agitation of a part or of the whole of the tongue, when the apparatus, formed of a still greater number of pairs of the said metals, is more active, and if, by means of good communicating conductors, the electric current which it excites be able to pass every where with perfect freedom.

I must often recur to, and insist on, this last condition, because it is essential in all experiments when you wish to obtain sensible effects on the body, or commotions in the limbs, or sensations in the organs of the senses. It is necessary, therefore, that the non-metallic conductors which enter into the circle should be as good conductors as possible, well moistened (if they are not themselves liquid) with water, or with any other liquid that may be a better conductor than pure water; and it is necessary, besides, that the well moistened surfaces, by which they communicate with the metallic conductor, should be sufficiently large. The communication ought to be confined or reduced to a small number of points of contact only in that place where you wish to concentrate the electric action on one of the most sensible parts of the body, on any of the sensitive nerves, &c. as I have already remarked in speaking of the experiments on feeling, *viz.* those by which acute pains are excited in different parts. The best method which I have found for producing on the tongue all the sensations above described, is, to apply the tip of it to the pointed extremity (which, however, must not be too much so) of a metallic rod, which I make to communicate properly, as in the other experiments, with one of the extremities of my apparatus, and to establish a good communication between the hand, or, what is better, both the hands together, and the other extremity. This application of the tip of the tongue to the end of the metallic rod, may either exist already, when you are going to make the other communication to complete the circle (when you are going to immerse your hand into the water of the bason), or be made after the establishment of this communication, while the hand is immersed; and in the latter case I think I feel the pricking and shock in the tongue, a very short time before actual contact. Yes; it always appears to me, particularly if I advance the tip of my tongue gradually, that, when it has arrived within a very small distance of the metal, the electric fluid (I would almost say spark), overcoming this interval, darts forwards to strike it.

In regard to the sense of sight, which I also found might be affected by the weak current of the electric fluid, arising from the mutual contact of two different metals in general, and in particular, of a piece of silver and one of zinc, it was natural to expect that the sensation of light, excited by my new apparatus, would be stronger in proportion as it contained a greater number of pieces of these metals; each pair of which, arranged in the proper manner, adds a degree of force to the said electric current, as all the other experiments show, and particularly those with the electrometer assisted by the condenser, which I have only mentioned, and which I shall describe on another occasion. But I was surprised to find that, with 10, 20, 30 pairs, and more, the flash produced neither appeared longer and more extended, nor much brighter than with one pair. It is true, however, that this sensation of weak and transient light, is excited by such an apparatus much easier and in different ways. To succeed, indeed, with one pair, the following are almost the only methods; *viz.* that one of the metallic pieces should be applied to the ball of the eye, or the eye-lid well moistened, and that it should be made to touch the other metal applied to the other eye, or held in the mouth, which produces a a flash much more beautiful; or, that this second metallic piece should be held in the moistened hand and then brought into contact with the former; or, in the last place, that these two plates should be applied to certain parts of the inside of the mouth, making them communicate with each other. But with an apparatus of 20 or 30 pairs, &c. the same flash will be produced by applying the end of a metallic plate or rod, placed in communication with one of the extremities of the apparatus, to the eye, while with one hand you form a proper communication with the other extremity; by bringing, I say, this plate into contact not only with the eye or any part of the mouth, but even the forehead, the nose, the cheeks, lips, chin, and even the throat; in a word, every part and point of the visage, which must only be well moistened before they are applied to the metallic plate. The form as well as the force of this transient light which is perceived varies a little, if the places of the face to which the action of the electric current is applied, be varied: if it be on the forehead, for example, this light is moderately bright, and appears like a luminous circle, under which figure it presents itself also in several other experiments.

But the most curious of all these experiments is, to hold the metallic plate between the lips, and in contact with the tip of the tongue; since, when you afterwards complete the circle in the proper manner, you excite at once, if the apparatus be sufficiently large and in good order, and the electric current sufficiently strong and in good order, a sensation of light in the eyes, a convulsion in the lips, and even in the tongue, and a painful prick at the tip of it, followed by a sensation of taste.

I have now only to say a few words on hearing. This sense, which I had in vain tried to excite with only two metallic plates, though the most active of all the *exciters* of electricity, *viz.* one of silver or gold, and the other of zinc, I was at length able to affect it with my new apparatus, composed of 30 or 40 pairs of these metals. I introduced, a considerable way into both ears, two probes or metallic rods with their ends rounded, and I made them to communicate immediately with both extremities of the apparatus. At the moment when the circle was thus completed I received a shock in the head, and some moments after (the communication continuing without any interruption) I began to hear a sound, or rather noise, in the ears, which I cannot well define: it was a kind of crackling with shocks, as if some paste or tenacious matter had been boiling. This noise continued incessantly, and without increasing, all the time that the circle was complete, &c. The disagreeable sensation, and which I apprehended might be dangerous, of the shock in the brain, prevented me from repeating this experiment.

There still remains the sense of smelling, which I have hitherto tried in vain with my apparatus. The electric fluid, which, when made to flow in a current in a complete circle of conductors, produces in the limbs and parts of the living body effects correspondent to their excitability, which stimulating in particular the organs or nerves of touch, taste, sight, and hearing, excite in them some sensations peculiar to each of these senses, as I have found, produces in the interior of the nose only a pricking more or less painful, and commotions more or less extensive, according as the said current is weaker or stronger. And whence comes it, then, that it does not excite any sensation of smell, though, as appears, it stimulates the nerves of that sense? It cannot be said that the electric fluid of itself is not proper for producing odorous sensations, since, when it diffuses itself through the air in the form of aigrettes, &c. in the common experiments made with electric machines, it conveys to the nose a very sensible smell resembling that of phosphorus. Taking similitude into consideration, and reasoning from its analogy with other odoriferous matters, I will say, that it must completely diffuse itself throughout the air to excite smell; that it has need, like other effluvia, of the vehicle of the air to affect that sense in such a manner as to excite the sensations of smell. But in the experiments of which I speak, that is to say, of an electric current in a circle of conductors, all contiguous, and without the least interruption, this absolutely cannot take place.

All the facts which I have related in this long paper in regard to the action which the electric fluid excited, and when moved by my apparatus, exercises on the different parts of our body which the current attacks and passes through; – an action which is not momentaneous, but which lasts, and is maintained during the whole time that this current can follow the chain not interrupted in its communications; in a word, an action the effects of which vary according to the different degrees of excitability in the parts, as has been seen; – all these facts, sufficiently numerous, and others which may be still discovered by multiplying and varying the experiments of this kind, will open a very wide field for reflection, and of views, not only curious, but particularly interesting to medicine. There will be a great deal to occupy the anatomist, the physiologist, and the practitioner.

It is well known, by the anatomy which has been made of it, that the electric organ of the torpedo or electric eel, consists of several membranaceous columns, filled from one end to the other with a great number of plates or pellicles, in the form of very thin disks, placed one upon the other, or supported at very small distances by intervals, into which, as appears, some liquor flows. But we cannot suppose that any of these laminæ are of an insulating nature, like glass, resin, silk, &c. and still less that they can either become electric by friction, or be disposed and charged in the same manner as the small Franklinian plates or small electrophores; nor even that they are sufficiently bad conductors to perform the office of a good and durable condenser, as Mr. Nicholson has supposed. The hypothesis of this learned and laborious philosopher, by which he makes of each pair of these pellicles, which he compares to leaves of talc, as many small *electrophores* or *condensers,* is indeed very ingenious, and is, perhaps, the best theory that has been devised to explain the phenomena of the torpedo, adhering to the hitherto known principles and laws of electricity. For, besides that the mechanism, by which, every time that the fish intended to give a shock, the respective separation of the plates of the whole or a great number of these electrophores or condensers ought to be effected all at once, and ought to establish on the one hand a communication between themselves of all the plates electrified *positively,* and on the other a communication between all those electrified *negatively,* as Mr. Nicholson supposes – besides, that this very complex mechanism appears too difficult, and little agreeable to nature; – and besides, that the supposition of an electric charge originally impressed, and so durable in these pellicles performing the office of electrophores, is altogether gratuitous, – such a hypothesis falls entirely, since these pellicles of the organ of the torpedo are not, and cannot be, in any manner insulating or susceptible of a real electric charge, and much less capable of retaining it. Every animal substance, as long as it is fresh, surrounded with juices, and more or less succulent of itself, is a very good conductor. I say more, instead of being as *cohibent* as resins or talc, to leaves of which Mr. Nicholson has compared the pellicles in question, there is not, as I have assured myself, any living or fresh animal substance which is not a better *deferent* than water, except only grease and some oily humours. But neither these humours nor grease, especially semi-fluid or entirely fluid, as it is found in living animals, can receive an electric charge in the manner of insulating plates, and retain it: besides, we do not find that the pellicles and humours of the organ of the torpedo are greasy or oily. This organ therefore, composed entirely of conducting substances, cannot be compared either to the electrophore or condenser, or to the Leyden flask, or any machine excitable by friction or by any other means capable of electrifying insulating bodies, which before my discoveries were always believed to be the only ones originally electric.

To what electricity then, or to what instrument ought the organ of the torpedo or electric eel, &c. to be compared? To that which I have constructed according to the new principle of electricity, discovered by me some years ago, and which my successive experiments, particularly those with which I am at present engaged, have so well confirmed, *viz.* that conductors are also, in certain cases, exciters of electricity in the case of the mutual contact of those of different kinds, &c. in that apparatus which I have named the *artificial electric organ,* and which being at bottom the same as the natural organ of the torpedo, resembles it also in its form, as I have advanced.

Source Endnote

1 Translated from the author's paper published in French in the Philosophical Transactions for 1800, part 2.

Editorial Headnote

16. "Mr. Bain's Electro Magnetic Inventions," *The Mechanics Magazine* 1041, July 22, 1843, pp. 64–79.

The Mechanics Magazine, Museum, Register, Journal, and Gazette, not to be confused with an American publication, *Mechanics Magazine,* was created by the London Mechanic's Institution to be "the most valuable gift which the hand of Science has yet offered to the artizan [sic]." The editors, along with many other education movements in the early nineteenth century, hoped to diffuse the history and principles of manufacture, science, and trade in a cheap and readable format. The pages frequently featured explanations of the works of noteworthy tinkerers and inventors like James Watt and Michael Faraday. They also contained explanations of prominent papers or descriptions of inventions. For example, here we follow the works of Mr. Alexander Bain (1810–1877). Bain apprenticed as a clockmaker in Edinburgh and later with John Barwise Jr. of the famed Barwise Watch and Clockmakers. He is most famous for his work on telegraphy and patent of the electric clock, which resulted in the feud with Charles Wheatstone mentioned in the article. He frequently delivered talks around Great Britain at the Polytechnic Institution and Adelaide Gallery. As he stated in his *A Short History of the Electric Clocks*, "I have devoted myself to rendering electricity practically useful." In this article, the author explores how Bain applied moving power to clocks and timepieces using electromagnetic force and applied electricity to control railways and carriages, mark time, give signals, and print intelligences at distant places. Beyond highlighting the workings of these devices and celebrating their utility, the author delves into the controversy between Bain and his partners in order to show that despite his illustrious reputation and career, Wheatstone was the "pretender" in this controversy. The essay covers letters, direct quotes, and other publications concerning this debate. Beyond exemplifying how widely and publicly scientific debates circulated during the nineteenth century, this essay highlights the broad network of people involved in the discovery and implementation of new electric technologies. Even though the feud took place between Wheatstone and Bain, the article included the input of William Cooke (1806–1879), John Frederic Daniell (1790–1845), Henry Moseley (1801–1872), Robert Willis (1800–1875), W. Snow Harris (1791–1867), George Birkbeck (1776–1841), and John Finlaison (1783–1860). Although they represent the elite echelons of British academic society, they make clear the vast networks that blossomed around electrical science between tinkerers, mathematicians, chemists, and natural philosophers. Furthermore, they appear here in the *Mechanics Magazine*, which appealed to a much wider audience.

16

MR. BAIN'S ELECTRO MAGNETIC INVENTIONS

J. C. Robertson

Source: J. C. Robertson ed., "Mr. Bain's Electro Magnetic Inventions," *The Mechanics Magazine* 1041, July 22, 1843, pp. 64–79.

On the 11th of January, 1841, Mr. Alexander Bain, a young watchmaker then recently arrived in London from Caithness, in conjunction with Mr. Barwise, the eminent chronometer maker of Saint Martin's-lane, obtained letters patent for "certain improvements in the application of moving power to clocks and time-pieces;" the moving power referred to being electro magnetism.

On the 21st of December following, Mr. Bain, and Lieut. Thomas Wright, R.N., took out another patent for "certain improvements in applying electricity to control railways and carriages, to mark time, to give signals, and print intelligence at distant places."

The two inventions, which form the subjects of these patents are popularly known by the names of the *Electro Magnetic Clock,* and *Electro Magnetic Printing Telegragh,* and have been made known to all the world by their exhibition during the last two years at that place of universal resort for the curious in scientific and mechanical matters, the Polytechnic Institution.

That by virtue of the above patents, the legal property of the Electro Magnetic Clock is vested in Messrs. Bain and Barwise, and of the Electro Magnetic Telegraph in Messrs. Bain and Wright has not been disputed by any one; and as long as they remain unrepealed, these parties must, and will be regarded in the eye of the law as respectively the only real inventors. It has been contended, however, on the part of Professor Wheatstone, the associate of Mr. W. F. Cooke in the invention of the Electric Telegraph, that whatever the law of the case may be, he was, in point of fact, the original author of both inventions; and gentlemen of high standing in the scientific world – Professors Daniell, Moseley, and Willis, Dr. Arnot and Mr. W. Snow Harris – have publicly certified, that in their opinion, Mr. Wheatstone's pretensions are well founded. The patentees say, on the other hand, that both the fact and the law are on their side, – that they are the exclusive proprietors of the disputed inventions by right of conscience as well as by royal license – and that Mr. Wheatstone is falsely and ungenerously seeking to arrogate to himself discoveries to which he has not the smallest claim. So far, however, – indeed, until the

DOI: 10.4324/9781003295556-19

very moment at which we are now writing – Mr. Wheatstone most he allowed to have had the best of it; for whether it be owing to the genuine strength of his case, or (as alleged) to the factitious influence of his position in society, it is certain, that he has hitherto carried the majority of the scientific world along with him, and that for the present the shadow of suspicion rests rather on the humble watchmaker of Caithness, than on his distinguished competitor of the King's College.

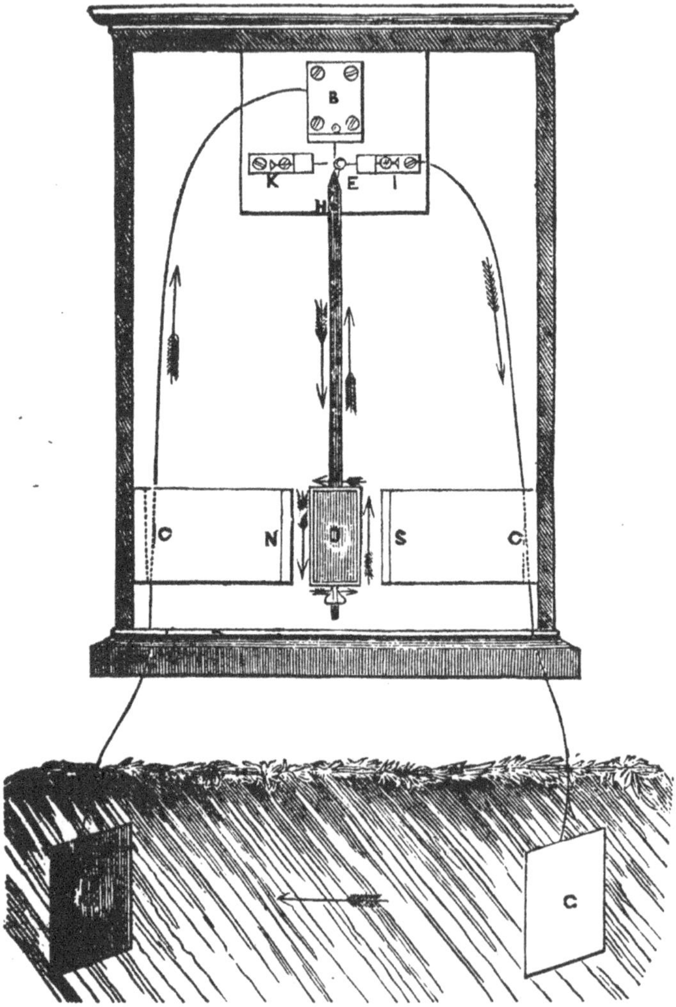

In this state of matters there has come to the rescue of Mr. Bain, and his co-patentees a champion of might enough to turn the tables against even greater odds – one, who, though inferior, doubtless, in scientific attainments to such men as Daniell, and Moseley, and Willis, and Arnott, will be allowed by all who know any thing of him to be able to cope with any of them on a question (such as this is) of "Probabilities" – a gentleman of great eminence in a most difficult branch of study, enjoying deservedly the confidence of government and the public, and distinguished beyond most men for habits of patient enquiry, and independent judgment; we mean, Mr. Finlaison, the Actuary of the National Debt Office and Government Calculator. A pamphlet from the pen of this gentleman, the title of which is subjoined,[1] has just made its appearance, in which, after a most acute and searching investigation of the rival claims to the invention of the Electro Magnetic Clock, and Electro Magnetic Printing Telegraph, he determines the question in favour of Mr. Bain and his friends, Messrs. Barwise and Wright. We are not insensible to some grave defects in the pamphlet; it is elaborate, and reiterative, for example, to a degree of weariness, and some very small matters are made a great deal too much of; but its defects are such as belong to the manner rather than the matter of the author; they tend to no wrong conclusions; they but render the way a little more devious – more thorny perhaps – to the one great and important conclusion at which he arrives, that Mr. Bain is the true man, and Mr. Wheatstone (so far as regards the inventions in question,) the pretender. For our own parts, we adopt this conclusion of Mr. Finlaison's with great pain – holding in very high esteem Mr. Wheatstone's unquestioned and unquestionable merits in many other respects; but at the same time, without hesitation; for it is a conclusion irresistibly forced upon us, by an attentive, dispassionate, and wholely unbiassed consideration of the facts and arguments adduced by Mr. Finlaison in its support.

The author's statement of the manner of his introduction to the acquaintance of Mr. Bain, and of his reasons for undertaking the task of his vindication, reflects great honour on both parties; and is interesting besides for the striking illustration which it contains of the importance of one of the inventions in dispute.

"It chanced that early in the last year, the writer, when inspecting the curiosities of the Polytechnic Institution, had occcasion to see with surprise and gratification, the action of a machine for conveying messages to far distant places by electrical agency, in such a manner that one person sitting in London could, with the speed of light itself, converse with another at Portsmouth, each delivering to his correspondent the precise words that he might wish to say, in print, on a scroll of paper, so that no mistake in the meaning could by possibility occur. Having passed the first seventeen years of his service at the Admiralty, in no very inferior station, the author ventures to think himself as well qualified as most people, to judge of the vast importance of such an invention to the interests of this great maritime nation.

"The ordinary semaphores are useless by night, and in our cloudy atmosphere are ineffective for nearly one-third of the days in the year. They are, moreover, erected in the first instance, and subsequently maintained at very great expense.[2] But here is a small portable engine, costing something like 100*l*., which can be set to work for less, perhaps than 50*l*. a mile in the outset, and worked afterwards for a trifling yearly salary, absolutely infallible in every state of the weather, by night or by day, in conveying words and sentences instantaneously and with perfect precision.

"Supposing, for instance, a time of war with France: in less than ten minutes such a message as the following would certainly reach the Admiralty office, either from Portsmouth or Plymouth:

> "The 'Prince Albert' is in sight and working in. She signals that the French fleet, yesterday at IV. P.M. put to sea from Brest, steering S.E. by S."

"The Lords of the Admiralty, after due deliberation, could cause such an order as the under-mentioned, to reach each of the senior officers at Sheerness, Deal, Portsmouth, and Plymouth, *simultaneously,* and infallibly within twenty minutes after it is written down at the Board:

> "The Brest fleet is at sea, probably XV sail of the line. Send all your disposable ships to rendezvous at Cork. The senior officer, when sufficiently in force, is to cruize in the chops of the channel, sending out steamers along shore to reconnoiter."

"No one acquainted with naval affairs can be ignorant that a time may arrive when the facilities afforded by such an agency for immediate defence, might be of the very utmost importance to the public safety.

"On enquiring who was the inventor of this extraordinary apparatus, the writer found, to his great surprise, that he was a young man, by name Alexander Bain, by trade a clock and watchmaker; a self-taught genius from the author's own native spot in the extreme North of Scotland, totally unfriended, and hitherto unknown to fame.

"There was further shown at the same institution, another of his inventions, consisting of a clock moved by electricity, which not only requires no winding up, but is capable of making any number (say 500) of other clocks simultaneously work together, by the one original movement. It may as well now be added, that, by a more recent discovery, Mr. Bain can make such a system of clocks keep time together for very many years, without their requiring any other trouble than such, perhaps, as an annual inspection." * * * * *

"At his young friend's request, the writer undertakes the task of stating his case for several reasons. The first is, that he cultivated with assiduity the pleasure of Mr. Bain's intimate acquaintance. Not only did he find him to be far above most others of his station in every praiseworthy quality – kind-hearted, sober, temperate, candid, incapable of deceit, and utterly free of guile; but, considering his years and opportunities, very well versed also in electrical science. Add to this, that in practical mechanics he is not merely an artist of the very first order (an immense advantage this over the man of books, because his hand can execute whatsoever his mind conceives), but he is endowed moreover with inventive faculties of such fertility that no one can see the exercise of his ingenuity without wondering what new invention may next occur to his prolific imagination. His last valuable discovery[3] was the result of extensive and repeated experiments made in the author's own grounds, and in his presence."

The chief grounds on which Mr. Wheatstone claims for himself the inventions in question, as stated by himself in two letters to the *Literary Gazette,* of the 18th June and 10th August, 1842, are these:

1. That, on the 26th Nov., 1840, he exhibited and described the electro magnetic clock to the Royal Society; which was before the thing was made public by any one else.
2. That the electro magnetic printing telegraph was but a direct and immediate application of the electro-magnetic telegraph invented by him in 1839, and patented 21st January, 1840.
3. That he had, in May, June, and July, of 1840, spoken about making such instruments to persons worthy of credit, who vouch for the fact.

And 4. That Mr. Bain, with whom he became first acquainted in August, 1840, was but a "working mechanic, who had been employed by him between the months of August and December of the year 1840," (which includes the period when the communication was made to the Royal Society), and "under a written engagement not to communicate what he was about to any other person without his (Mr. W.'s) permission;" whence the reader is evidently desired to infer that all Mr. Bain's pretensions to the character of an independent inventor are founded on a base betrayal of his employer's confidence.

Mr. Finlaison admits it to be clearly enough proved that Mr. Wheatstone had, previous to his acquaintance with Mr. Bain, entertained the idea of applying electric currents to the moving of clocks, and also of causing the electric telegraph to print, as well as exhibit, words and signals at a distance; but he submits a counter statement of facts which go to prove that, beyond *talking* about the practicability of such things, (a very different thing from actually carrying them out into practice,) Mr. Wheatstone really did nothing until *after* he became acquainted with Mr. Bain, and until *after* Mr. Bain had shown and left with him models both of an electro magnetic clock and electro magnetic telegraph which he had *before* constructed, and shown to other parties; and that the connexion between Mr. Wheatstone and Mr. Bain was not as the former represents it to have been, that of master and servant, but simply that of a humble and meritorious inventor in search of a liberal and honest patron. We insert, in an abridged form, as much of Mr. Finlaison's narrative, as may serve to put the reader in possession of the leading points of the case: –

> "Mr. Bain, in the exercise of his vocation as a clock and watch-maker, arrived in London at the age of twenty-six, in March, 1837. It is proved by a letter from Mr. Charles M'Dowall, watchmaker, then of St. James's-street, now of Beaufort-street, Chelsea, that as early as March, 1838, Mr. Bain had told him of his intention to apply the electric fluid to the movement of clocks, '*so as to make any number go together.*' No man will venture to assert that the idea of *measuring time,* so as to make any

number of clocks go together by electricity, had occurred to any other person than Mr. Bain at this early date, although in the previous year (and in one instance as long as twenty years ago) many others had thought of applying the electric fluid to telegraphic communications, on account of its property of passing with the speed of light to unknown distances. In the course of the two following years Mr. Bain had not only matured his invention of the system of electro magnetic clocks, as above, but he had extended his application of the same agency to various other purposes; especially to the transmission of telegraphic messages in such a form as that the message should be delivered IN PRINT to those for whom it was intended. It is certified by the same Mr. M'Dowall, that in the course of the month of June, 1840, he personally inspected Mr. Bain's model of the electro magnetic clock, and also 'the apparatus *for printing at a distance by means of electricity*;' and that he further saw 'several pieces of paper which had been printed on by the apparatus.' It is also certified, in a letter from Mr. Robert C. Pinkerton, of 20, Perceval-street, Clerkenwell, watchmaker, and therefore another competent judge, that early in the month of July, 1840, he too inspected Mr. Bain's model of his electro magnetic printing telegraph, with which Mr. Bain printed his (Mr. Pinkerton's) name. He certifies, moreover, that, at the same visit, Mr. Bain showed him 'a model of the electro magnetic clock, and explained to him the principles and utility of them.' No man will hazard the assertion, that up to these dates the electro magnetic power had ever actually performed the work of telegraphic printing, in any mode or fashion, by the hands of any other person than those of Mr. Bain."

* * * *

"By the month of June, 1840, Mr. Bain was prepared with models of each of the two inventions in question, to be laid before men of science and capital, in the hope that these would promote the introduction of them into practical use. To this end he waited on Alderman Sir Peter Laurie, early in the morning of the 1st of August following, who gave him an introductory letter to Dr. Birkbeck.

"With this letter Mr. Bain immediately waited on Dr. Birkbeck, who was then, however, unfortunately, too ill to be seen. Not knowing in that case what to do, he repaired to the office of the *Mechanics' Magazine,* in Fleet-street, which he reached about noon, *all on the same day* [*quod nota bene*]. Here he met with Mr. Baddeley, who was then the assistant editor of that journal, to whom he minutely explained his two inventions, and who recommended him to wait on Professor Wheatstone, with the same explanations, at King's College, in the Strand, close by, with which Mr. Bain *immediately* complied, and saw the Professor for the first time on that very morning.

"At this, the first interview, the Professor listened to Mr. Bain's disclosures with profound attention and apparent surprise. But giving no hint that he was himself then occupied with similar projects, he said that he would see him again on the subject at another time, and fixed their second interview at his own house, in Conduit-street, for the 18th of August. They met accordingly at the Professor's house, when Mr. Bain produced his models, viz., the one that of the electric clock, the other that of the printing telegraph. The Professor advised Mr. Bain *to postpone, for the time being,* all proceedings with reference to the electric clock, without the smallest hint that he had ever thought of such a thing himself, but agreed to purchase the model of the printing telegraph, disclosing, for the first time, that he also was then contriving a printing telegraph on another construction, but which contrivance he only explained in part; but nothing in the shape of a model of it was ever shown. Some scrap of a diagram, on a little bit of paper, and that only of a small portion of it, was indeed lying on the table.

Two papers were *drawn up by the Professor* on the spot, and signed by them respectively. Mr. Bain understood that they were both to the same effect. That in his possession is as follows: –

> "'In addition to the sum of five pounds, which I have given you for the *model* of your proposed modification of the printing apparatus, to be added to the electric telegraph, I will give you fifty pounds, in the event of my making any profitable application thereof; that is, should I have any instrument made for sale in which the inking roller is employed, and in which the wheel, on the circumference of which the types are placed, is bodily moved forward, in order to impress the types on the cylinder carrying the paper, instead of the types being pressed individually as in my instrument.
>
> "'(Signed) C. WHEATSTONE.'

"'London, 20, Conduit-street,
August 18*th*, 1840.'

"In the paper retained by the Professor, there was originally inserted a proviso, to the effect that Mr. Bain should not dispose of any model of a printing telegraph to any other person. But Mr. Bain decidedly objected to this clause, having in contemplation a very important extension of the powers of that engine beyond what it was capable of in the simpler model thus disposed of. The Professor readily acquiesced in the objection and struck out the passage. By this obliteration, then, the real counterpart will be known, if Mr. Wheatstone shall produce it. The Professor dares not assert that any other written agreement ever passed between them. At this point of time, Mr. Bain considered their connexion at an end: he had sold, and Mr. Wheatstone had purchased on certain conditions, the model of one of his inventions in the state in which it then stood, declining to purchase the other, viz., that of the electro magnectic clock, and exhorting Mr. Bain very earnestly not to proceed any further in the working out this invention. This advice, however, Mr. Bain disregarded, as he not only proceeded forthwith to construct another printing telegraph on far more elaborate principles, and of greater power; but also to look out for a friend to join him in patenting the invention of the clock, which patent was actually applied for on the 10th October, just fifty-three days after this interview. It would have been sealed in a few weeks more, but for the interesting situation of Her Majesty, whose first confinement took place on the 24th November, and who could not therefore be troubled to affix the royal sign manual to the warrant for sealing.

* * * * *

"During the month of September, succeeding the sale of his model to Mr. Wheatstone, and while his patent for the electric clock was in progress of negociation, Mr. Bain turned his attention to that improvement on his first invention of the printing telegraph, which he had in contemplation at the time he sold the model of it, as already mentioned, the object of which improvement was chiefly to produce greater rapidity in printing the message, though by more complex mechanism than the former and simpler machine was capable of effecting. Not doubting that this improved principle would be highly acceptable to the Professor, he waited upon him, for the third time, at his own house, with a rough model of this new invention. But Mr. Wheatstone objected to the much greater expense which would attend the new principle, as well as the increased wear and tear consequent on the added mechanism, and the danger of the apparatus falling more frequently into disorder. He ended by declining to have any concern with it.

* * * * *

"[In consequence of a subsequent call which Mr. Wheatstone made on Mr. Bain,] they entered into a new agreement in September, 1840, into which the former one was to merge, but without the precaution of committing it to writing; so unlimited at that time was his reliance on the honour of Mr. Wheatstone. By the new treaty it was provided, that the professor should from time to time supply the necessary funds for the materials and workmanship of two perfect working models, one of the simpler, the other of the more complex printing telegraph, which Mr. Bain had invented; and that on the application of either of those principles to the intended improved telegraph of Mr. Wheatstone, Mr. Bain should receive a further sum of £150 as the price of both inventions. It cost the Mecænas nothing to make a promise which he could at any time annul, as in fact he did. He had before then, but on what precise date cannot at present be discovered, slily set one of his artificers to work on the construction of a clock, which should be a perfect *facsimile* of that of Mr. Bain. Pursuant to the new treaty, Mr. Bain's labours proceeded for several months; the working model of the telegraph on the simpler principle was finished – that on the more complex plan was far advanced. But the time had arrived for throwing him overboard, the Professor's clock being now secretly finished. Mr. Bain had received, in various small sums, altogether about £25, and the first symptoms of estrangement which he discovered were, that the supplies began to be more and more reluctantly yielded. On being importuned, the Professor desired the finished model to be taken to him on a given day at King's College, with which Mr. Bain complied, in the hope of obtaining some part of the payment now due to him. At the time appointed, however, the Professor was invisible, but had left orders with the porter to take in the model, which was left accordingly. Although, thus disappointed of the needful advances, Mr.

Bain continued his labours on the second or more complex model, until, at length, he found that the money in arrear amounted to as much as he had received. Becoming impatient, he was desired to carry the second model, in its unfinished state, to the Professor's house for inspection. This he complied with, and also left the model there, having evoked a small instalment of £3. But he never again set eyes on either the first finished model, nor on this the second unfinished one, nor yet on his well-earned money. Ten times over did Mr. Bain call for the unfinished work in order to resume his labour, and ten times over was the Professor denied to him. At last, although the most unsuspicious of human beings, Mr. Bain began to fancy that his models were in some sort purloined, and that he himself was little better than a dupe. About the middle of December, 1840, determined to have an explanation with Mr. Wheatstone, he called at his house at the unusual hour of eight o'clock in the evening, and had his final interview. To his utter astonishment, the Professor put himself into a violent passion, the cause of which Mr. Bain could at first by no means comprehend. Being reproached, however, and in a tone of '*Non tibi sunt integra linlea,*' with the inutility of his inventions, and that they were inferior to those of the Professor himself, Mr. Bain at once saw that he was not only to be choused, but bullied, and he retorted accordingly. Mr. Bain, however, was the first to recover his decorum, and retired from this disgraceful scene, he being *minus* and the Professor *plus* his models and £25 of arrears, to say nothing of his further expectation of £150 eventually.

* * * * *

"Mr. Wheatstone (Nov. 26, 1840) exhibited the clock to the Royal Society with a memoir, in which he claims the exclusive merit of the invention. It was afterwards shown to the Society of Civil Engineers, and its properties explained by Mr. Dent, the eminent chronometer-maker of the Strand. The same artist attended the British Association, subsequently, at their annual meeting which, in 1841, was at Plymouth, and exhibited the clock, with ample details of its manifold merits, as the invention of Mr. Wheatstone, which he really believed. The proceedings of those illustrious bodies were, however, far beyond the ken of such a retired student as Mr. Bain; but one day in January, 1841, Mr. Barwise having accidentally learned that this pretended invention was to be exhibited that same evening at the Adelaide Gallery, was just in time to serve Mr. Wheatstone, and Mr. Dent, one of the directors of that institution, with notice of an injunction. Thus the Professor's model was perforce withdrawn, and Mr. Bain's clock very soon afterwards exhibited at the Polytechnic."

Hard as this statement bears on Mr. Wheatstone, it has a great air of truth about it; it is established in all its material points by the documentary evidence referred to; and is simple and consistent throughout. The endeavour of Mr. Wheatstone to make it appear that Mr. Bain was but "a working mechanic, who had been employed" by him, under an obligation of secrecy, which he violated, tells more against the learned professor, in our opinion, than any one thing in the whole case. Most certainly there was no such relationship as that of master and servant between the parties – no such obligation of secrecy; and what else can we conclude, but that it is a bad cause which is attempted to be supported by such false pretexts? It is manifest that while Mr. Wheatstone was but thinking of making a clock and (printing) telegraph, Mr. Bain had, *without any communication with Mr. Wheatstone,* actually constructed and shown models of both; and Mr. Wheatstone does not pretend to have exhibited any model of either till *after* he had seen Mr. Bain and his models, and *after* Mr. Bain (in conjunction with Mr. Barwise) had commenced the taking out of a patent for the clock. We have had some experience in patent cases, and we say unhesitatingly that, were it attempted, on such a state of facts, to rescind by *scire facias* the patent of Messrs. Bain and Barwise, the verdict must be for the defendants. As Sir Isaac Newton – an authority Mr. Wheatstone will not, we presume, be inclined to dispute – said of Leibnitz with respect to the discovery of fluxions, "Whether he had the notion from me, or invented it after me, is of no consequence – *as second inventors have no rights.*"

It is true that in the memorandum of the 18th August, 1840, which Mr. Wheatstone gave to Mr. Bain, he speaks of "*my* instrument" for printing, as a thing then invented and in existence; but to know what value ought to be attached to this circumstance, we must view it in connexion with certain other circumstances, disclosed in the following additional extract from Mr. Finlaison's statement: –

> "He (Mr. Wheatstone) does not pretend that any one ever saw complete drawings, or any model or other apparatus for the purpose, until the month of January, 1842, when the specification of his next patent was enrolled. His evidence of such prior contrivance is only the admission of Mr. Bain, on the 18th of August (in the agreement which he himself drew up for Mr. Bain to sign) that the idea was then on the anvil. That it was no farther advanced is proved by his own witness, as follows:
>
> " 'St. Petersburg Place, Bayswater,
> 29th July, 1842.
>
> " 'Dear Sir, – At the time you mentioned to me that you had contrived an addition to your electric telegraph, by which it could be made to print the letters, instead of merely *showing them,* you *asked me for some information respecting the mode of preparing the manifold writing paper, which you proposed to employ, and on the best form of type for obtaining impressions with it.* The note, in which I answered these inquiries respecting your printing telegraph, was dated June 10th, 1840.
>
> " 'I remain, dear Sir, yours sincerely,
>
> " '(Signed) EDWARD COWPER.'
>
> " 'Professor Wheatstone, F.R.S.,' &c.

"Thus it is manifest that the notion of printing, though it may have been entertained by the Professor, was inchoate on the 10th of June, just fifty-one days before Mr. Bain explained how he was at that moment performing the act of printing with engines ready for the purpose. Mr. Pinkerton states, that early in July, 1840, he saw his own name printed by Mr. Bain; and Mr. M'Dowall saw the apparatus for printing and its work in the previous month, viz., at the same instant of time when the Professor was only consulting with Mr. Cowper how to set about such printing. Nothing daunted, however, the Professor concludes his first letter (to the *Literary Gazette* of the 18th June) by saying, 'Whatever may be the merits of Mr. Bain's method, it cannot justify any person to call in question the originality *and priority* of my electro magnetic *printing* telegraph, which is secured from infringement by two patents, one already mentioned, including the telegraph itself, and the other of more recent date, comprising my superadded printing apparatus.' Thus it is admitted that the patent 'already mentioned,' being that specified on the 21st July, 1840, merely *showed* letters, as Mr. Cowper truly states; but never was designed to *print* them. The 'other of more recent date' is that specified on 7th January, 1842, and is the first in which the invention of printing is mentioned. The reader needs scarcely to be informed that the specification of a patent is the formal publication of the various objects to which the owner by that act acquires a property, and the contrivances by which these are to be attained. Before the sealing of the patent, which in this case occurred on the 7th July, 1841, it was necessary to lodge with the Attorney-General, under the seal of secrecy, a briefer specification, known by the name '*Deposit,*' and which sets forth each and every object of the patent in general terms. Such a document was lodged by Mr. Wheatstone with Sir John Campbell, who then held that high office, on 9th June, 1841. Soon after this the ministry was changed, and Sir Frederick Pollock succeeded.

"In the meantime, Mr. Bain, having greatly improved his former invention of the printing telegraph, had associated with Lieutenant Thomas Wright, of the Royal Navy, for the purpose of obtaining a patent for the engine in this its improved state. To their infinite surprise, they were vehemently opposed by Mr. Wheatstone in person, before the new Attorney General, on 6th October, 1841. Sir Frederick Pollock having been left in utter ignorance by that ingenuous philosopher, that he had lodged with Sir John Campbell, on the 9th June preceding, the deposit of what he calls his electro magnetic *printing telegraph,* and being entirely persuaded by the professor that he, and he alone, was the inventor of telegraphic printing, absolutely refused the same privilege to Messrs. Bain and Wright. They retired in great disappointment; but having accidentally heard of the deposit, they, in a few days, wrote to Sir Frederick Pollock, and prayed him to unseal that paper. The officer of the crown might very naturally be somewhat annoyed at perceiving he was imposed upon by something like a trick. He examined the deposit accordingly, and the result was, that he immediately passed Mr. Bain's patent. The whole of this strange mystification is explained in the following letter, in answer to an inquiry made by Lieutenant Wright: –

"'Temple, Feb. 1, 1843.

"'Sir, – The Attorney General has opened and read the paper deposited by Professor Wheatstone, on the 9th June, 1841; *and he finds no mention made of a 'printing telegraph.'* I lose no time in making this communication to you.

" 'I remain, Sir, your humble servant,

(Signed) " 'Jos. Coleman.' "

"Thos. Wright, Esq."

"Now the professor is here placed on the horns of a dilemma. If the mere insertion of printing in his specification, lodged on 7th January, 1842, notwithstanding the non-insertion of any such operation in his original deposit of June preceding – printing not being, therefore, within the purview of the patent when sealed in July, 1841 – be nevertheless sufficient to convey to him, as he says, the exclusive property in that invention, it was equally competent to him to have inserted the right of printing in his specification of the signalizing telegraph, which was lodged on 21st July, 1840, ten days before he saw Mr. Bain, in which case at this earlier date he could have unquestionably secured the invention by law. But as no such right was ever claimed, the inference is inevitable – *that up to his first interview with Mr. Bain, he had not, in fact, matured any practical contrivance for the purpose, notwithstanding his reiterated assurances to the contrary.* But if the Professor shall perch himself on the other horn, and say, that although he had, *before* he saw Mr. Bain, contrived '*far more efficient means*' than he for printing, but was deterred from the insertion of those means in his specification of 21st July, 1840, by reason that printing was not in the purview of the patent so specified, and could not therefore be protected by it, then it results, of necessity, that his patent 'of more recent date, comprising his superadded *printing apparatus*' is, as far as printing goes, not worth a farthing."

The more Mr. Wheatstone's pretensions are enquired into, the more he suffers by the enquiry; for the present, therefore, we shall take our leave of him, in order to bring under the notice of our readers some further advances made by Mr. Bain in the application of the electric fluid to the useful arts, as we find them described in the pamphlet before us.

In our Journal of the 11th of June, 1842, we gave an account of some remarkable experiments which we had witnessed on the Serpentine River, made by Mr. Bain and Mr. Wright, in which a voltaic circuit had been accomplished, not by a continuous series of metal wires, as usual, but partly by wires and partly by the water of the river. The following is Mr. Bain's account of the experiment, as well as of some others which preceded and followed it.

"In prosecuting some experiments with an electro magnetic sounding apparatus in the year 1841, it was found that if the conducting wires were not perfectly insulated from the water in which they were immersed, the attractive power of the electro magnet did not entirely cease when the circuit was broken. For the purpose of investigating the nature of this phenomenon, a series of experiments took place with great lengths of wire, in the reservoir of water at the Polytechnic Institution, when similar results were obtained.

"As this effect threatened to militate against the practical application of electric currents to the transmission of telegraphic communications or the working of electric clocks, it became a matter of considerable importance to discover, if possible, the true cause of the phenomenon, and to provide a remedy. With this view, therefore, Lieut. Wright and Mr. Alexander Bain, having obtained the permission of His late Royal Highness the Duke of Sussex for that purpose, commenced a series of extensive and varied experiments on the Serpentine river, in Hyde-park, which it has been thought necessary to particularize, showing as they do the steps which have led to one of the most extraordinary, and, it would seem, one of the most valuable discoveries yet made in electrical science.

"The first experiment consisted in passing an electric current through the water, by means of wires laid from one side of the river to the other, as shown in the *diagram, experiment* 1.

"*A* represents a compound battery of six cells, of about twelve square inches surface; *B* and *C* are the wires; *D* is an electro magnet of soft iron, and *E* its feeder. On an electric current being established in the wires, it was found that a small portion only reached the electro magnet: enough, however, to enable it to sustain its own weight. On the circuit being broken, by disconnecting the wires *from the battery,* it was found that the attractive power of the magnet did not entirely cease. The electric current being again transmitted through the wires, the circuit was broken by detaching the wires *from the magnet,* when its attractive power ceased immediately. The experiment was then repeated as at first, and the same result obtained – viz., a very gradual decay of the magnetic power. It is well to observe here, that the feeder was removed from the magnet, and kept from it several minutes; on being again presented to the magnet, it was slightly attracted by it. It was premised that on an electric current being established in one direction, its effect on the magnet might be instantly annihilated, by changing the direction of the current; experiment proved this to be the case, and thus pointed out an effectual remedy for the inconvenience, although the cause was still unknown. As it was evident in the foregoing experiment that the greater portion of the electricity was conducted *from one wire to the other by the water,* particular attention was next given to this branch of the subject. A portion of one of the wires forming the circuit, was lifted out of the water at several points between the two banks of the river, and the electro magnet placed in the circuit; when it was found that the current was transmitted by the water from one wire to the other, as shown by the small arrows: the greatest portion of the electric current passing from that part of the wires which was nearest to the battery.

"These facts rendered it obvious that water was quite capable of conducting voltaic electricity, provided a sufficient surface of metal was present to convey the current into and out of the water. Before proceeding to apply this fact, however, the first experiment was repeated, but with a smaller battery. See *experiment* 2, *in diagram.* – *F* is a small Grove's battery of four inches surface; *G* and *H* the positive and negative wires; *I* the galvanometer. The results of this experiment were exactly similar to the former, except that a larger proportion of the diminished current reached the galvanometer, as was anticipated from the metallic surface of the wires being an ample conductor for so small a current.

"A copper wire was next laid down on the gravel walk along the north bank of the Serpentine River, from the bridge which separates Hyde Park from Kensington gardens, to the east end of the river. About three square feet of metallic surface was

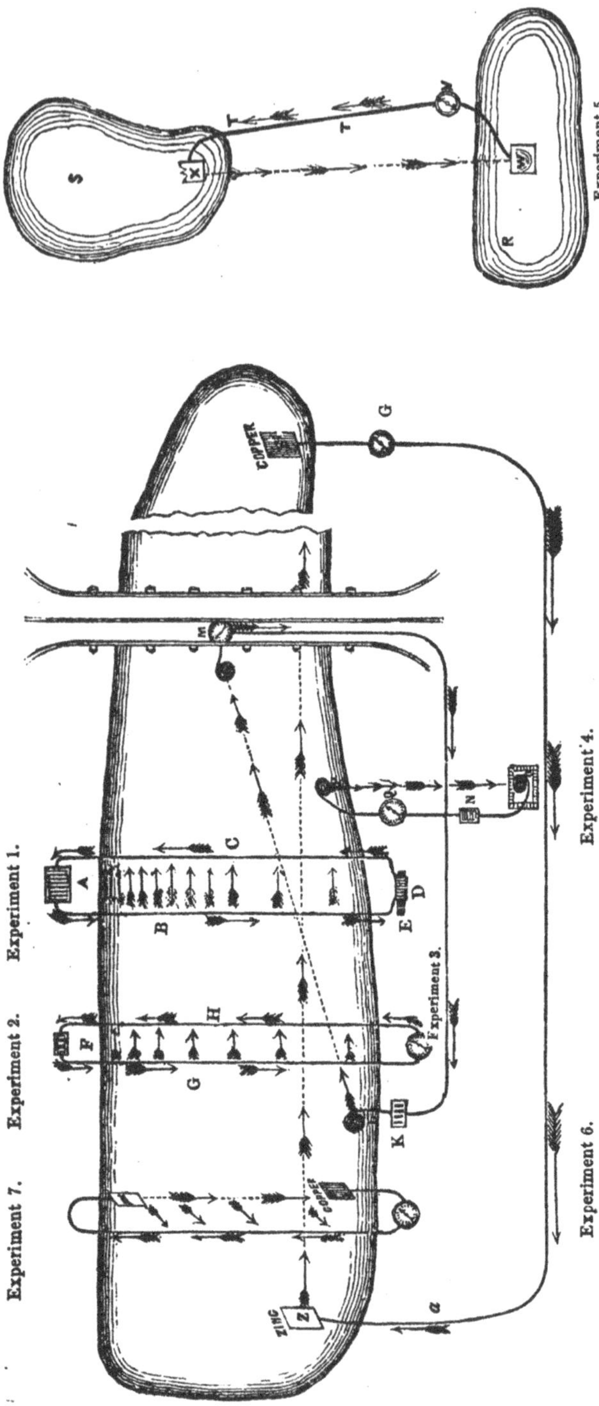
Experiment 5
Experiment 1.
Experiment 2.
Experiment 7.
Experiment 3.
Experiment 4.
Experiment 6.
COPPER
ZINC
COPPER

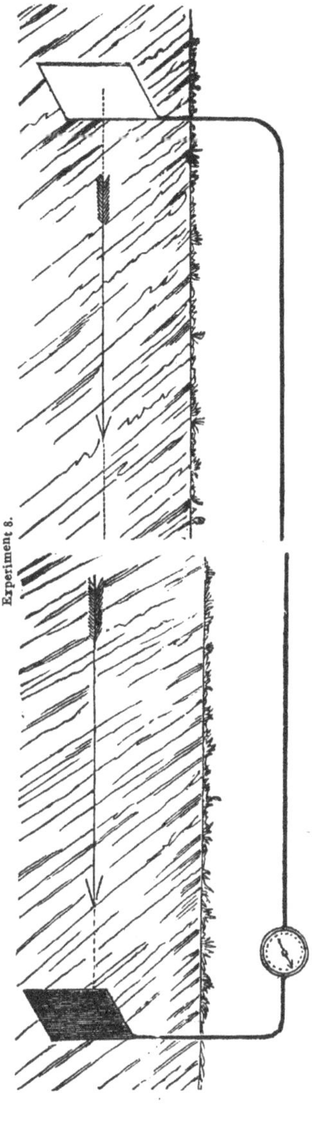

Experiment 8.

attached to each end of the wire and put into the river; a galvanometer was put into the circuit at the bridge, and the small Grove's battery at the other extremity of the wire.

"See *diagram, experiment* 3. – *K* is the battery, *L L* the conducting wires, M the galvanometer. The electric current passed by the water and returned by the wire, as shown by the arrows, and with as much power as would have been the case with an ordinary metallic circuit.

"In this arrangement the magnetic influence ceased the moment the circuit was broken, as it would have done in an entirely dry circuit.

"Reflecting on the foregoing experiment, it occurred to Mr. Bain that the natural moisture of the earth might be a sufficient conductor for the electric current, and with a view to ascertain the correctness of this assumption, a wire was led along a wood paling extending from the river to a well about a hundred and fifty yards distant; one of the metallic surfaces attached to the wire was put into the river, and the other into the well; the galvanometer was put into the circuit near the river, and the small battery near the well.

"See *diagram, experiment* 4. – *N* is the battery, *O* the metallic surface in the well, *P* that in the river, and *Q* the galvanometer. On completing the circuit, the current passed freely, as in the former experiments, showing that *when sufficient moisture* is present, the earth is a good conductor of voltaic electricity; and that one-half of a voltaic circuit is all that is necessary to be insulated from surrounding conducting matter.

"While reflecting upon these experiments, some few months after they had been' performed, Mr. Bain was led to infer, that if a surface of postitive metal was attached to one end of a conducting wire, and an equal surface of negative metal to the other end, and the two metallic surfaces put into water, or into the moist earth (the wire being properly insulated from surrounding matter,) an electric current of considerable energy would be established in the wire.

"This proposition was soon tested by an experiment performed in the grounds of Mr. Finlaison, (the government calculator,) at Algher's house, Loughton, Epping-forest. See *diagram, experiment* 5. – *R* is a moat, distant 150 yards from *S,* a pond of water, *T T* a conducting wire laid along a gravel walk, *V* a galvanometer, *W* represents about 12 inches surface of positive metal, and *X* a similar quantity of negative metal.

"The moment this arrangement was completed, the galvanometer showed that an electric current was passing from the metal in the pond, through the earth, in the direction of the arrows, to the metal in the moat, returning back again by the wire *T.* The current was of considerable energy, and this experiment was repeated a number of times with unvarying success.

"From observations that had been made daring the prosecution of the foregoing experiments, it was next resolved to try the effect of a water battery upon a large scale. This was done in the Serpentine river, as shown in the *diagram, experiment* 6. A large surface of copper, *Y,* being placed in the water, a corresponding surface of zink was inserted at *Z:* the two surfaces being connected by the wire *a a a.* On placing a galvanometer in the circuit at *G,* an electric current of considerable intensity was found to be passing through the water from the zink to the copper, and returning by the wire. This experiment was varied, by passing the current across the Serpentine with the wire immersed in the water, as shown in the *diagram, experiment* 7. *Z,* as before, is the zink and *Y* the copper, *G* the galvanometer. With this arrangement, a portion of the electricity was transmitted from one side of the river to the other, but the greater part of the current from the plates passed through the water to the wire *en route.*

"These points being satisfactorily established, Mr. Bain next proceeded to make the experiment as shown in the *diagram,* No. 8.

"A surface of zink was buried in the moist earth in Hyde Park, and, at rather more than a mile distance, a copper surface was similarly deposited; the two metals were connected by a wire, suspended on the railing, and on placing a galvanometer in the circuit, an electric current was produced, which passed through the intervening mass of earth from one plate to the other, returning by the wire. In the first experiment, the metallic surfaces being small, the electric current produced was feeble, but on using a larger surface of metal a corresponding increase in the energy of the current was obtained, with which an electrotype process was conducted, and various electro magnetic experiments performed with uniform success.

"Subsequent experiments have shown that if two metal plates (a negative and positive) of sufficient surface are sunk in the earth as a battery, and wires led therefrom, electrotype deposition may be effected, and every description of electro magnetic apparatus worked for any length of time. The most successful results have, however, been obtained by depositing several surfaces of positive metal in the earth, connected into a group by wires, from which a conducting wire was led to a series of negative surfaces similarly disposed at a more remote spot. When considerable power is required, this is the arrangement that should be adopted. It is essential to success, that the earth wherein the plates of metal are deposited should be of a moist nature. A current has, indeed, been obtained in dry soils, but of such small energy as to be of no practical utility. This, however, may have been occasioned by the very small proportion of metallic surface with which the experiment was made.

"Such a source of electricity as the foregoing promises to be most extensively useful in the arts. Among other advantages, its simplicity and cheapness are no small recommendations; while the uniform character of its power is of the utmost importance. A battery of this description, under very disadvantageous circumstances, has produced a power which for upwards of six months has been found unvarying."

Mr. Finlaison insists very earnestly on the originality of this discovery, and we must freely confess that when we witnessed the experiments at the Serpentine, the results they furnished were new to us. We know that other persons, too, much more conversant with electrical science than we can pretend to be, thought that the results were new. Mr. Wheatstone himself, for example – who *repeated* Messrs. Bain and Wright's experiments by establishing an electric communication between the King's College, and the lead-shot tower on the opposite side of the Thames, by means of wires carried from the College, along the parapets of Somerset House and Waterloo Bridge, and thence to the tower, and Surrey side of the river, whence the electric current was transmitted across by the water; and who subsequently gave an account of this feat to the Institution of Civil Engineers, but (with *characteristic candour*) forgot to take any notice of what Messrs. Bain and Wright had previously accomplished in the same way. All these things, however, notwithstanding, there is an extract given from Mr. Sturgeon's valuable Annals of Electricity in a letter which precedes this article, from Mr. Cooke, the partner of Mr. Wheatstone in some of his inventions, which, it must be admitted, puts the claim to novelty quite out of court; for it shows that as early as 1838 Professor Steinheil had distinctly pointed out that an electric circuit might be formed partly of metallic conductors, and partly of earth or water. All that can therefore be fairly conceded to Messrs. Bain and Wright is, the merit of having re-discovered this fact, and brought prominently under public notice the important purposes to which it may be turned. On this latter point Mr. Finlaison indulges in some anticipations too startling almost for belief; but if the fact of the transmission of an electric current through such a body of water as the Thames or Serpentine be once admitted, (and who can dispute it?) it is evident that this is a case in which we must believe a great deal more than it is given to us (as yet) to understand.

"That the water or moisture of the earth would, to some extent, conduct voltaic electricity, is a fact which might have been known in the scientific world, without any useful application of it. To this end a further stretch of thought became necessary. Mr. Bain found that if the mere ends of the wire were dipped into the water, the current that passed was so feeble, that if he had stopped here, it would be applicable to no practical use. He then saw that it was necessary to attach a few feet of metallic surface to be immersed at each end of the insulated conducting wire. The result was, that the whole current of the small battery employed immediately passed as freely through the water as it could have done through an entire metallic circuit. The success of that experiment obviously led Mr. Bain to his great discovery of plunging, in a similar manner, plates of positive and negative metallic surfaces in the earth, or in the water, at great distances, then connecting these by a well insulated wire, he is enabled to discard galvanic batteries altogether, and to produce an everlasting and unvarying flow to of electricity, proportionate in power the amount of metallic surface. The importance of such a discovery cannot be overrated, and it is hoped that Mr. Wheatstone will now at last be pleased to leave Mr. Bain quietly in possession also of the honour. If a copper wire, one-sixth of an inch in thickness, be imbedded in a bar of boiling asphaltum, and sent along the railway (for its better protection) from London to Liverpool – if two tons weight of zink plates be immersed in the Mersey at Liverpool, and attached to that end of the wire – and if one ton weight of copper be sunk in the river Thames, and attached to this end of the wire, no rational man can doubt that an electric current would be established of ten times the power necessary to work a telegraph, What is 200 miles, or any terrestrial distance, to an agency that travels with the speed of light? By two independent astronomical phenomena, namely, the aberration of light, and the retardation of the eclipses of Jupiter's satellites, it is alike established that this influence travels, in round numbers, at the rate of 192,000 miles per second. Mr. Wheatstone says, that he has discovered the velocity of electricity to be still greater, and gives, as his conclusion, that it goes at the rate of 197,000 miles per second. Now, in the voltaic circuit in question, it may be a hard matter to say whether the current flies round the coast by the sea, or whether it penetrates the earth superficially. One thing is however certain, the experiment is not likely to be tried at the sole cost of the inventor – and it is fairly to be conceived that the country which is to be benefited by the discovery should incur the trifling charge of bringing it into use. Supposing, however, the very improbable event that the current, from some unforeseen cause, could not be passed to the distance of 200 miles, it may still be sent as far as it will go on railways, connecting these very easily by means which are well known to every practical electrician." – pp. 33, 35.

We hope, though as it were hoping against hope, that the appeal here made to the wisdom and generosity of the country will be responded to.

Mr. Bain has made a very ingenious application of his re-discovered fact of the electric conducting power of the earth to the improvement of his electro-magnetic clock, and this, at least, we take to be perfectly new. The engraving, page 65, exhibits what Mr. Bain calls his latest improved pendulum, and the following is Mr. Finlaison's description of it.

"*A A* is a mahogany case with a glass front; *B* is a metal bracket fixed to the back of the case, and to which the pendulum *D* is suspended. *C C* are permanent steel magnets fixed to the sides of the case in such a manner as that the pendulum-ball *D* can vibrate freely between the poles of each magnet. The magnets are so placed as that poles of dissimilar names face each other. *E* is a small platinum ball affixed to a brass stem, free to move to one side or the other, being fastened to a light spindle carried by the pendulum-rod at *H.* The plate of copper *F* is deposited in the moist earth, from which a wire leads to the bracket *B.* The plate of zink *G* is likewise deposited in the earth, and its wire leads to the piece of metal *I.* To the lower end of the suspension-spring of the pendulum is attached a wire coated with silk. It is led down the back of the rod (which is wood,) and then coiled longitudinally, in many convolutions, around the edge of the pendulum-ball, in a groove previously made for the purpose. It is then taken up the back of the rod and terminates in the bearings of the spindle at *H.* The action of the engine is as follows: – A constant and uniform current of electricity would be established, and would pass through the earth, the plates and wires in the direction of the arrows, as long as the platinum ball *E* rests on the platinum pin projecting from the metal *I.* But if the pendulum is put in motion, suppose that, at first, it were drawn aside until the ball *D* should be between the poles of the right-hand magnet, the point *H* being now farther to the right than the ball *E,* the latter would fall to the left and rest on the pin *K* until the pendulum took its vibration to the left, when the ball *E* would fall to the right, and so on continually, the action being produced by the change of the centre of gravitation at each vibration of the pendulum. This action of the ball *E* lets on and cuts off the flow of electricity at or near the extreme ends of the pendulum's vibrations, so that the convolving wire of the pendulum-ball is attracted and repelled by the magnets at the proper points of its vibrations, and thus a continual motion is kept up for an indefinite period of time."

Here the friction of the moving parts is diminished to a degree which it seems impossible for human agency to exceed, and in proportion to that diminution must of course be the accuracy of the instrument as a measurer of time. Mr. Finlaison regards it with great show of reason as being probably the nearest approach yet made to the impossible chimera called the perpetual motion.

STIRLING'S AIR MACHINE

"Sir, – I went to see Mr. Stirling's air-engine at work yesterday. I saw one at work last year; he has since altered the steam-engine used for driving the machinery in his foundry, and it works very beautifully indeed. He states the consumption of fuel to be only one-fourth of what it was. He has tried the power with a friction break, and ascertained the consumption per horse power, so that there is 'no mistake about it.' His plan of heating and condensing the air is, I think, most beautiful and ingenious. The beaters occupy very little space, and stand close to the engine. He has, and I think very properly, been working it this long time to gain experience, before introducing it to the world; and is now about to alter another steam-engine used to drive a spinning-mill in town.

"N."

"Dundee, July 8, 1843."

SHIPS' PUMPS

"Sir, – In a letter from Mr. Halse, suggesting improvements in ships' pumps, inserted in your Number of the 17th ult., are one or two errors, which, with your leave, I will here refer to. In the first place, he is quite mistaken in supposing the *Conqueror* and the *Reliance* were lost through the inefficiency of their pumps. These vessels were lost through their running on shore, and knocking themselves to pieces: the best pumps in the world would have been of no use to them: *nothing* could have saved them but getting afloat. Supposing it possible to draw the water from the hold of a leaky vessel, by means of a syphon, no doubt it would be very desirable; but, if I understand the principle of a syphon, the one thing necessary is, to have the *discharging mouth below the level of the liquid to be discharged,* a condition in this case impossible to be fulfilled. A syphon might certainly be used (in smooth water) for the second purpose, viz., to draw water from the sea to the hold of the vessel; but here, also, the discharging mouth must be below the liquid to be removed; therefore the water could only be sent to the *lower* part of the hold.

"I am, Sir, yours truly,

"T. W."

A Return of the original Cost of the Engines of the following Steam Vessels, specifying any extra charge beyond Contract Price, and if such Vessels were fitted with an Indicator; of Repairs, and the Cost of such Repairs, and the number of Days and Hours any of said Vessels were incapable of performing their work, in consequence of such Repairs; and of the names of the Makers of each of the Engines of said Vessels: –

Names of Vessels.	Original Cost of Engines.	Extra Charge beyond Contract price.	Indicator.	REPAIRS.			Cost of Repairs.			Number of Days and Hours the Vessels were incapable of performing their Work, in consequence of such Repairs.	Names of the Makers of the Engines.
				Nature	Period over which they extend. From	To					
	£.	£.					£	s.	d.		
Alecto. . . .	10,700	297	not fitted	{ two engines, boilers and paddle-wheels.}	13 Jan. 1840	31Mar. 1843	1,158	4	4¾	393	Messrs. Seaward and Capel.
Devastation	18,650	681	ditto	ditto	Dec. 1841	ditto	249	10	8½	92	Messrs. Maudslay, Son, and Field.
Geyser. . . .	13,933	440	ditto	ditto	July 1842	ditto	89	7	3½	50	Messrs. Seaward and Capel.
Cyclops. . .	22,103	906	ditto	ditto	Oct. 1840	ditto	800	7	10	164	ditto
Prometheus	10,700	315	ditto	ditto	March1840	ditto	1,012	1	5½	353	ditto
Polyphemus	10,700	214	ditto	ditto	June 1841	ditto	240	7	11¼	162	ditto
Vesuvius. . .	13,480	400	ditto	ditto	Oct. 1840	ditto	37	19	0¼	38	Mr. Robert Napier.
Stromboli	13,480	nil.	ditto	ditto	Sept. 1840	ditto	68	6	3¼	51	ditto

Admiralty, 26th May, 1843.

WILLIAM EDWARD PARRY,
Comptroller of Steam; Machinery.

A RETURN of the Names of Marine Engine-Makers with whom the Admiralty have made Contracts for Engines, from the year 1839 to 1843 inclusive; with the amount of Horse-Power ordered from each of such Engine-Makers in each year: –

Name of Engine-Makers.	Amount of Horse-Power in the Years				
	1839.	1840.	1841.	1842.	1843.
Mr. John Laird.	265	–	–	–	–
Messrs. Maudslay, Son, and Field . .	200	400	420	2,230	–
Messrs. Seaward and Capel	800	1,400		1,035	–
Messrs. Miller and Ravenhill	. .	. .	286	430	–
Messrs. Fawcett and Co.	300	. .		350	–
Messrs. Fairbairn and Co.	. .	300	20	470	–
Mr. Robert Napier.	. .	. .	300	–	–
Messrs. Boulton and Watt.	. .	. .	300	–	–
Messrs. G. and J. Rennie.	. .	. .		450	–
Messrs. Scott and Sinclair.	. .	. .	280	–	–
Mr. Robinson.	. .	. .		220	–
Messrs. Acraman and Morgan	. .	. .	20	–	–
Messrs. Penn and Son.	. .	. .		260	–

WM. EDWARD PARRY,
Admiralty, 26th May, 1843. Comptroller of Steam Machinery.

Source Endnotes

1 An Account of some Remarkable Applications of the Electric Fluid to the Useful Arts, by Alexander Bain. With a Vindication of his claim to be the First Inventor of the Electro Magnetic Printing Telegraph, and also of the Electro Magnetic Clock. By John Finlaison, Esq., Actuary of the National Debt Office, and Government Calculator, 128 pp., 8vo., with numerous plates. Chapman and Hall, 1848.

2 See a Return printed by order of the House of Commons, on 5th May, 1843, No. 236, from which it results that the semaphore communicating between the Admiralty and Portsmouth, is only available for 100 hours out of every 545, and that the expense of working it is, on an average of three years, at the rate of 3,403*l*. 9*s*. 6*d*. per annum, exclusive of the first cost of erection and repairs, and exclusive also of the ground which it was necessary to purchase in the first instance.

3 The passing of the electric current through the earth without the aid of any galvanic battery. See *post.*

Editorial Headnote

17. "On the Production of Insects by Voltaic Electricity" from a letter from Richard Phillips to William Sturgeon, *The Annals of Electricity, Magnetism, and Chemistry*, 1 April 1837, pp. 242–244.

Correspondence provides historians of science with an incredible lens into the lives and work of nineteenth-century naturalists, natural philosophers, and scientists. Massive collections of these correspondences have been collected and annotated in the last decade; see the Darwin or Tyndall Correspondence Projects. Letter writing offered a medium through which vast networks of scholars might communicate prior to international meetings and societies. Frequently, letters functioned as first drafts and review for later publications and experience. Beyond working as a window into scientific operations, letters can provide sociocultural context and even bring less-published scholars to the forefront of construction of scientific knowledge at the time. Here we see a letter from the publisher Richard Phillips (1767–1840) to the "electrician" William Sturgeon (1783–1850) concerning an experiment by Andrew Crosse (1784–1855) and requesting that Sturgeon might place the letter in the upcoming journal. Sturgeon founded the *Annals of Electricity* in 1836 and published it until 1843. Building on his work on exhibitions, public lectures, and experimentation, he believed a publication dedicated exclusively to electricity was necessary. Crosse gained fame for his research on electricity and crystals, or electro-crystallization, which drew attention from the British Association for the Advancement of Science. Additionally, he studied electric phenomenon in the atmosphere and the creation of voltaic piles. However, he is perhaps more notorious for the two experiments printed here in the *Annals*. In the first experiment, Crosse poured a highly conductive fluid into a basin in which he suspended a piece of porous volcanic rock, attached the rock by wire to a voltaic battery, and kept the rock saturated with the fluid so that it was "kept constantly electrified." After two weeks, he observed minute white specks. Over the next few days, he recorded the growth of hundreds of animalcules that he said resembled cheese mites. In the second experiment, he completed a similar process, but instead observed the mite growing on the surface of the fluid. He readily admits to not being an entomologist and welcomes additional experimentation. He makes it clear that he did not think the electric action created life. In his memoirs, published by his wife Cornelia A. H. Crosse, the experiment was "in friendly spirit" published in a newspaper, and he flew around the country drawing scorn for taking the role of creator. His *Acarus galvanicus* had "greatness thrust upon them." Even in a publication concerned with explaining new technologies such as the galvanometers, electric power still captured wonder and highlighted anxieties about the potential of electric science.

17

ON THE PRODUCTION OF INSECTS BY VOLTAIC ELECTRICITY

Andrew Crosse

Source: Andrew Crosse, "On the Production of Insects by Voltaic Electricity" from a letter from Richard Phillips to William Sturgeon, *The Annals of Electricity, Magnetism, and Chemistry*, 1 April 1837, pp. 242–244.

XLVI. *On the production of Insects by Voltaic Electricity. By* ANDREW CROSSE, ESQ., &c, &c. *Communicated in a letter from* SIR RICHARD PHILLIPS, *to* W. STURGEON, ESQ. &c. &c.

Dear Sir,

This morning I received the enclosed from Mr. Crosse, of Broomfield. Its contents are so novel and interesting that I am persuaded you will, even at this eleventh hour, give it place in your forthcoming number. With my best wishes for the success of your valuable Journal,

I am, respectfully,
R. PHILLIPS.

Cadogan Place,
March 25th. 1837.

The following is the account of two experiments in which the birth of Insects was the unexpected result.

Broomfield, March 23d. 1837.

Dear Sir,

Experiment 1. I poured a dilute solution of silicate of potash, super-saturated with muriatic acid, into a quart basin standing on a piece of mahogany in a wedgwood funnel, and caused the fluid to fall by successive drops into the funnel, by means of a strip of flannel wetted with the solution which acted as a syphon, and was suspended over the edge of the basin. These drops were made to fall upon a piece of somewhat porous red oxide of iron from Vesuvius, which was supported by a glass funnel which conveyed the fluid into a bottle below, and which, when full, was emptied back into the basin above, without disturbing the position of the stone. Two platina wires at the opposite ends of the stone, were connected with

DOI: 10.4324/9781003295556-20

the positive and negative poles of a voltaic battery, consisting of nineteen pairs of five-inch plates in cells filled with water and muriatic acid. Thus the stone was kept constantly electrified by the battery, and moist by the dropping. At the end of fourteen days, two or three very minute white specks or nipples were visible on the surface of the stone, between the two wires, by means of a lens. On the eighteenth day these nipples elongated and were covered with fine filaments. On the twenty-second day their size and elongation increased, and on the twenty-sixth day each figure assumed the form of a perfect insect, standing on a few bristles which formed its tail. On the twenty-eighth day these insects moved their legs, and in the course of a few days more, detached themselves from the stone, and moved over its surface at pleasure, although in general they appeared averse to motion, more particularly when first born. In the course of a few weeks more, about a hundred of these insects made their appearance on the stone, I observed that each of them fixed itself for a considerable time in one place, as far as I could judge appearing to feed by suction; but when a ray of light from the sun was let fall upon it, it seemed much disturbed, and finally removed itself to the shaded side of the stone. I ought to state that out of about a hundred insects, not above six or seven were born on the south side of the stone. I examined some of these insects with the microscope. Their shape was similar to that of the cheese mite, but they seemed to be from twice to eight times the size of that animalcule, and the motion of their legs was occasionally discernible by the naked eye. They had in general eight legs, but there were some with only six; their bodies were covered with bristles which gave some of them the appearance of a little star; the length of the bristles at the tail was considerable, and these when highly magnified appeared spicated. Instant death followed their removal from the stone however delicately performed, but in the course of time most of them were washed down into the bottle below, and poured into the basin above, from whence they found their way to the wooden support which was kept moist by the fluid dropping between it and the wedgwood funnel. On examining this piece of wood I found upwards of forty living insects actively crawling over its surface, and apparently feeding on the silicious matter with which it was impregnated.

Experiment 2. I filled a small glass jar with a concentrated solution of silicate of potash, and plunged into it a thick iron wire connected with the positive pole of a battery composed of twenty pairs of cylinders in cells filled with common water, and immersed in the same a small coil of fine silver wire connected with the negative pole of the same battery. In the course of a few weeks gelatinous silex was formed in some quantity round the iron wire, and after a longer interval of time, a similar formation, but in much less quantity, filled up the silver coil. On examining these wires with a lens, I first observed a similar insect at the negative wire, and some time afterwards I found two more; in the whole, three at the negative, and shortly afterwards I found twelve at the positive; in the whole fifteen. Each of these insects was well formed, and quite as large as those observed in the first experiment. They were deeply imbedded in the gelatinous silex, with the bristles of the tail projecting. They were in general from half to three quarters of an inch below the surface of the fluid. In this experiment there was neither flannel, wood, volcanic stone, nor acid, consequently the insects appear to have been born in the silex. Not being an entomologist, I can give no opinion on the matter. I have only related facts. Many persons have seen the insects both dead and living. Whether the electric action has any thing to do with their birth or not I cannot say without further experimenting. Perhaps if these experiments were repeated and varied, curious facts might be the result. Since the above, having found that all the insects in the first experiment had died, I poured into the basin an additional quantity of the same fluid, and in the course of two or three weeks I had a second crop of insects upon the volcanic stone. I have also had two other formations in two distinct solutions of silicate of potash; also five others in the following solutions: 1, a weak solution of oxide of copper; 2, 3, 4, 5, in *concentrated* solutions of green sulphate of iron, sulphate of zinc, sulphate of copper, and *nitrate of copper* – in the whole, *ten* separate formations; *five* in silicious solutions, and five in solutions of metallic salts &c. – each of them was exposed to a long continued electric action, before any insect made its appearance, still without further experimenting I would not venture an opinion. I hope in the course of time to know something more about this. I am going on well in crystallization. I have had a very handsome letter from Mons. Becqueril.

Your's sincerely,

(To Sir Richard Phillips.) ANDREW CROSSE.

Editorial Headnote

18. *Endless Amusement: a collection of nearly 400 entertaining experiments in various branches of science* (Thomas Boys, Ludgate-Hill: London, 1825), pp. 88–89, 100.

As noted by Graeme Gooday, didactic writings on electricity frequented the publications of the Victorian world. As illustrated here and in many other monographs, late nineteenth-century authors produced volumes of texts on electricity in textbooks, novels, short stories, poetry, and periodicals for audiences that ranged from novice to expert. People agonized over the potential of electricity, they feared it, and sometimes they thought it the key to understanding the rest of nature. However, instructive writing on electricity also ventured into other spaces, such as books filled with experiments for children. As Tina Young Choi demonstrates, play functioned as means through which Victorians might learn to experience an increasingly uncertain world. Here you will see several pages from such a work on amusement published by the British publisher and printmaker Thomas Boys (fl. 1819–1853) in Ludgate Hill, London. Many others have written at length about the significance of moral and instructive tales in Victorian children's literature. This work falls into the latter category. Throughout the pages of this book, the reader could find instructions for everything from making homemade invisible ink to fireworks or a perpetual motion machine. The inclusion of the various recipes is clearly intended to entertain. Very few of the items in these pages are dedicated to enhancing the readers' understanding of the guiding theory behind the work; in fact, many of the descriptions might be best categorized as practical jokes. About halfway through the book, the author moves to describe a set of experiments that revolve around the manipulation of electric currents. It is worth noting that this was published during the 1820s, shortly after Michael Faraday (1791–1867) invented the electromagnetic motor and George Ohm (1789–1854) mathematically described an electric circuit. Therefore, many of the experiments described within the text were not only for entertainment but reflected major theoretical shifts in physics and chemistry. For instance, the author describes experiments that require the harnessing of electric power, knowledge of ferrous metals, and awareness of theories, such as the magnetic phenomenon that causes the Aurora Borealis. However, it is impossible to gauge the exact readership of this book or explicitly determine the author's sources. (He makes no mention of any natural philosophers in the book.) This book again highlights the range of audiences that publishers believed electricity (and other scientific principles) might attract. Electrical experiments and imaginaries did not extend solely to social commentary or scientific theory but also shaped entertainment and play for inexpert audiences. The usage of such sources should be encouraged among historians of science and technology seeking to better understand the reception of major changes and topics among the established scientific community.

18

ENDLESS AMUSEMENT

A collection of nearly 400 entertaining experiments in various branches of science

Anon

Source: Anon. *Endless Amusement: a collection of nearly 400 entertaining experiments in various branches of science* (Thomas Boys, Ludgate-Hill: London, 1825), pp. 88–89, 100.

The Electrical Fountain

SUSPEND a vessel of water from the middle of the brass arch, and place in the vessel a small tube. The water will be one continued stream; and if the electrification be strong, a number of streams will issue, in form of a cone, the top of which will be at the extremity of the tube. This experiment may be stopped and renewed almost instantly, as if at the word of command.

The Electric Kite

MAKE a small cross of two light strips of cedar, the arms so long as to reach to the four corners of a large thin silk handkerchief when extended; tie the comers of the handkerchief to the extremities of the cross; and you have the body of the kite, which being properly accommodated with a tail, loop, and string, will rise in the air like those made of paper; but this being of silk, is more adapted to bear the wet and wind of a thunder gust, without tearing. To the top of the upright stick of the cross is to be fixed a very sharp-pointed wire, rising a foot or more above the wood. To the end of the twine, is to be tied a silk ribbon, and where the silk and twine join, a key may be fastened. This kite is to be raised when a thunder-storm appears to be coming on; and the person who holds the string must stand within a door or window, or under some cover, so that the silk ribbon may not be wet; and care must be taken that the twine do not touch the frame of the door or window. As soon as any of the thunder clouds come over the kite, the pointed wire will draw the electric fire from them, and the kite, with all the twine, will be electrified, while the loose filaments of the twine will stand out every way, and be attracted by an approaching finger. When the rain has wetted the kite and twine,

DOI: 10.4324/9781003295556-21

so that it can conduct the electric fire freely, you will find it stream out plentifully from the key, on the approach of your knuckle. At this key an electric phial may be charged; and from electric fire thus obtained, spirits may be kindled, and all the other electric experiments performed, which are usually done by the help of a rubbed glass or tube, and thereby the identity of the electric matter with that of lightning completely demonstrated.

[. . .]

The Electrified Ball

PLACE an ivory ball on the prime conductor of the machine, and take a strong spark, or send the charge of a Leyden phial through its centre, and the ball will appear perfectly luminous: but if the charge be not sent through the centre, it will pass over the surface of the ball and singe it. A spark made to pass through a ball of box-wood, not only illuminates the whole, but makes it appear of a beautiful crimson, or rather fine scarlet colour.

Editorial Headnote

19. "On Electric Conduction and the Nature of Matter," a letter to Richard Taylor, *Experimental Researches in Electricity* vol. 2 (London: Richard and John Edward Taylor, 1844), pp. 284–293.

Michael Faraday (1791–1867) was and remains one of the most influential figures in the history of the early nineteenth-century electrical sciences. His biography is well-known. Most notably, he developed an electromagnetic motor, described electromagnetic induction, and, as shown in the following piece, speculated about the extension of electromagnetic forces in the "empty" space around conductors. Beyond his experimental prowess, Faraday became an icon in British science for his ascendance from a bookbinder's apprentice to an eminent lectureship at the Royal Institution and Royal Society as a gentleman scientist. Here, Faraday wrote to Richard Taylor (1781–1858), a naturalist best known for his work as an editor for the *Philosophical Magazine* and *Annals of Natural History*. In addition to describing his theory and experiment, Faraday wrote to Taylor about the possibility of publishing. This letter appears among several letters and papers published in a collection on experimental researches in electricity. This paper is not necessarily his most important paper, nor is it necessarily the theory to which he contributed the most. This paper is instead emblematic of the way that Faraday, and many other natural philosophers at the time, used electricity in their research. Certainly, they explored the phenomenon on its own, but they also used it to develop other areas of interest or express their thinking on scientific practice. For example, in this letter Faraday relayed his electrochemical work on the conductivity of metals and discusses his theories on magnetism. At the same time, as is made most clear by the opening of the letter, one of his main concerns was addressing the development of the understanding of the atomic constitution of matter. He believed his work on electromagnetism and conductivity might elucidate other natural philosophers on understanding of the "empty space" around atoms. Beyond addressing theory and explaining his own work, Faraday also saw this as a space to explore the meaning of hypothesis, facts, truth, and probability in scientific work.

19

ON ELECTRIC CONDUCTION AND THE NATURE OF MATTER

Michael Faraday

Source: Michael Faraday, "On Electric Conduction and the Nature of Matter," a letter to Richard Taylor, *Experimental Researches in Electricity* vol. 2 (London: Richard and John Edward Taylor, 1844), pp. 284–293.

A speculation touching Electric Conduction and the Nature of Matter[1]

To Richard Taylor, Esq

DEAR SIR, Royal Institution, Jan. 25, 1844.

LAST Friday I opened the weekly evening-meetings here by a subject of which the above was the title, and had no intention of publishing the matter further, but as it involves the consideration and application of a few of those main elements of natural knowledge, facts, I thought an account of its nature and intention might not be unacceptable to you, and would at the same time serve as the record of my opinion and views, as far as they are at present formed.

The view of the atomic constitution of matter which I think is most prevalent, is that which considers the atom as a something material having a certain volume, upon which those powers were impressed at the creation, which have given it, from that time to the present, the capability of constituting, when many atoms are congregated together into groups, the different substances whose effects and properties we observe. These, though grouped and held together by their powers, do not touch each other, but have intervening space, otherwise pressure or cold could not make a body contract into a smaller bulk, nor heat or tension make it larger; in liquids these atoms or particles are free to move about one another, and in vapours or gases they are also present, but removed very much further apart, though still related to each other by their powers.

The atomic doctrine is greatly used one way or another in this, our day, for the interpretation of phenomena, especially those of crystallography and chemistry, and is not so carefully distinguished from the facts, but that it often appears to him who stands in the position of student, as a statement of the facts themselves,

DOI: 10.4324/9781003295556-22

though it is at best but an assumption; of the truth of which we can assert nothing, whatever we may say or think of its probability. The word atom, which can never be used without involving much that is purely hypothetical, is often *intended* to be used to express a simple fact; but good as the intention is, I have not yet found a mind that did habitually separate it from its accompanying temptations; and there can be no doubt that the words definite proportions, equivalents, primes, &c., which did and do express fully all the *facts* of what is usually called the atomic theory in chemistry, were dismissed because they were not expressive enough, and did not say all that was in the mind of him who used the word atom in their stead; they did not express the hypothesis as well as the fact.

But it is always safe and philosophic to distinguish, as much as is in our power, fact from theory; the experience of past ages is sufficient to show us the wisdom of such a course; and considering the constant tendency of the mind to rest on an assumption, and, when it answers every present purpose, to forget that it is an assumption, we ought to remember that it, in such cases, becomes a prejudice, and inevitably interferes, more or less, with a clear-sighted judgment. I cannot doubt but that he who, as a wise philosopher, has most power of penetrating the secrets of nature, and guessing by hypothesis at her mode of working, will also be most careful, for his own safe progress and that of others, to distinguish that knowledge which consists of assumption, by which I mean theory and hypothesis, from that which is the knowledge of facts and laws; never raising the former to the dignity or authority of the latter, nor confusing the latter more than is inevitable with the former.

Light and electricity are two great and searching investigators of the molecular structure of bodies, and it was whilst considering the probable nature of conduction and insulation in bodies not decomposable by the electricity to which they were subject, and the relation of electricity to space contemplated as void of that which by the atomists is called matter, that considerations something like those which follow were presented to my mind.

If the view of the constitution of matter already referred to be assumed to be correct, and I may be allowed to speak of the particles of matter and of the space between them (in water, or in the vapour of water for instance) as two different things, then space must be taken as the only continuous part, for the particles are considered as separated by space from each other. Space will permeate all masses of matter in every direction like a net, except that in place of meshes it will form cells, isolating each atom from its neighbours, and itself only being continuous.

Then take the case of a piece of shell-lac, a non-conductor, and it would appear at once from such a view of its atomic constitution that space is an insulator, for if it were a conductor the shell-lac could not insulate, whatever might be the relation as to conducting power of its material atoms; the space would be like a fine metallic web penetrating it in every direction, just as we may imagine of a heap of siliceous sand having all its pores filled with water; or as we may consider of a stick of black wax, which, though it contains an infinity of particles of conducting charcoal diffused through every part of it, cannot conduct, because a non-conducting body (a resin) intervenes and separates them one from another, like the supposed space in the lac.

Next take the case of a metal, platinum or potassium, constituted, according to the atomic theory, in the same manner. The metal is a conductor; but how can this be, except space be a conductor? for it is the only continuous part of the metal, and the atoms not only do not touch (by the theory), but as we shall see presently, must be assumed to be a considerable way apart. Space therefore must be a conductor, or else the metals could not conduct, but would be in the situation of the black sealing-wax referred to a little while ago.

But if space be a conductor, how then can shell-lac, sulphur, &c. insulate? for space permeates them in every direction. Or if space be an insulator, how can a metal or other similar body conduct?

It would seem, therefore, that in accepting the ordinary atomic theory, space may be proved to be a non-conductor in non-conducting bodies, and a conductor in conducting bodies, but the reasoning ends in this, a subversion of that theory altogether; for if space be an insulator it cannot exist in conducting bodies, and if it be a conductor it cannot exist in insulating bodies. Any ground of reasoning which tends to such conclusions as these must in itself be false.

In connexion with such conclusions we may consider shortly what are the probabilities that present themselves to the mind, if the extension of the atomic theory which chemists have imagined, be applied in conjunction with the conducting powers of metals. If the specific gravity of the metals be divided by the atomic numbers, it gives us the number of atoms, upon the hypothesis, in equal bulks of the metals. In the following table the first column of figures expresses nearly the number of atoms in, and the second column of figures the conducting power of, equal volumes of the metals named.

Atoms.		Conducting power.
1·00	gold	6·00
1·00	silver	4·66
1·12	lead	0·52
1·30	tin	1·00
2·20	platinum	1·04
2·27	zinc	1·80
2·87	copper	6·33
2·90	iron	1·00

So here iron, which contains the greatest number of atoms in a given bulk, is the worst conductor excepting one; gold, which contains the fewest, is nearly the best conductor. Not that these conditions are in inverse proportions, for copper, which contains nearly as many atoms as iron, conducts better still than gold, and with above six times the power of iron. Lead, which contains more atoms than gold, has only about one-twelfth of its conducting power; lead, which is much heavier than tin and much lighter than platina, has only half the conducting power of either of these metals. And all this happens amongst substances which we are bound to consider, at present, as elementary or simple. Whichever way we consider the particles of matter and the space between them, and examine the assumed constitution of matter by this table, the results are full of perplexity.

Now let us take the case of potassium, a compact metallic substance with excellent conducting powers, its oxide or hydrate a non-conductor; it will supply us with some facts having very important bearings on the assumed atomic construction of matter.

When potassium is oxidized an atom of it combines with an atom of oxygen to form an atom of potassa, and an atom of potassa combines with an atom of water, consisting of two atoms of oxygen and hydrogen, to form an atom of hydrate of potassa, so that an atom of hydrate of potassa contains four elementary atoms. The specific gravity of potassium is 0·865, and its atomic weight 40; the specific gravity of cast hydrate of potassa, in such state of purity as I could obtain it, I found to be nearly 2, its atomic weight 57. From these, which may be taken as facts, the following strange conclusions flow. A piece of potassium contains less potassium than an equal piece of the potash formed by it and oxygen. We may cast into potassium oxygen atom for atom, and then again both oxygen and hydrogen in a twofold number of atoms, and yet, with all these additions, the matter shall become less and less, until it is not two-thirds of its original volume. If a given bulk of potassium contains 45 atoms, the same bulk of hydrate of potassa contains 70 atoms nearly *of the metal potassium,* and besides that, 210 atoms more of oxygen and hydrogen. In dealing with assumptions T must assume a little more for the sake of making any kind of statement; let me therefore assume that in the hydrate of potassa the atoms are all of one size and nearly touching each other, and that in a cubic inch of that substance there are 2800 elementary atoms of potassium, oxygen and hydrogen; take away 2100 atoms of oxygen and hydrogen, and the 700 atoms of potassium remaining will swell into more than a cubic inch and a half, and if we diminish the number until only those containable in a cubic inch remain, we shall have 430, or thereabout. So a space which can contain 2800 atoms, and amongst them 700 of potassium itself, is found to be entirely filled by 430 atoms of potassium as they exist in the ordinary state of that metal. Surely then, under the suppositions of the atomic theory, the atoms of potassium must be very far apart in the metal, *i. e.* there must be much more of space than of matter in that body: yet it is an excellent conductor, and so space must be a conductor; but then what becomes of shell-lac, sulphur, and all the insulators? for space must also by the theory exist in them.

Again, the volume which will contain 430 atoms of potassium, and nothing else, whilst in the state of metal, will, when that potassium is converted into nitre, contain very nearly the same number of atoms of potassium, *i. e.* 416, and also then seven times as many, or 2912 atoms of nitrogen and oxygen besides. In carbonate of potassa the space which will contain only the 430 atoms of potassium as metal, being entirely filled by it, will, after the conversion, contain 256 atoms more of potassium, making 686 atoms of that metal, and, in addition 2744 atoms of oxygen and carbon.

These and similar considerations might be extended through compounds of sodium and other bodies with results equally striking, and indeed still more so, when the relations of one substance, as oxygen or sulphur, with different bodies are brought into comparison.

I am not ignorant that the mind is most powerfully drawn by the phenomena of crystallization, chemistry and physics generally, to the acknowledgement of centres of force. I feel myself constrained, for the present hypothetically, to admit them, and cannot do without them, but I feel great difficulty in the conception of atoms of matter which in solids, fluids and vapours are supposed to be more or less apart from each other, with intervening space not occupied by atoms, and perceive great contradictions in the conclusions which flow from such a view.

If we must assume at all, as indeed in a branch of knowledge like the present we can hardly help it, then the safest course appears to be to assume as little as possible, and in that respect the atoms of Boscovich appear to me to have a great advantage over the more usual notion. His atoms, if I understand aright, are mere centres of forces or powers, not particles of matter, in which the powers themselves reside. If, in the ordinary view of atoms, we call the particle of matter away from the powers *a,* and the system of powers or forces in and around it *m,* then in Boscovich's theory *a* disappears, or is a mere mathematical point, whilst in the usual notion it is a little unchangeable, impenetrable piece of matter, and *m* is an atmosphere of force grouped around it.

In many of the hypothetical uses made of atoms, as in crystallography, chemistry, magnetism, &c., this difference in the assumption makes little or no alteration in the results, but in other cases, as of electric conduction, the nature of light, the manner in which bodies combine to produce compounds, the effects of forces, as heat or electricity, upon matter, the difference will be very great.

Thus, referring back to potassium, in which as a metal the atoms must, as we have seen, be, according to the usual view, very far apart from each other, how can we for a moment imagine that its conducting property belongs to it, any otherwise than as a consequence of the properties of the space, or as I have called it above, the *m*? so also its other properties in regard to light or magnetism, or solidity, or hardness, or specific gravity, must belong to it, in consequence of the properties or forces of the *m,* not those of the *a,* which, without the forces, is conceived of as having no powers. But then surely the *m* is the *matter* of the potassium, for where is there the least ground (except in a gratuitous assumption) for imagining a difference in kind between the nature of that space midway between the centres of two contiguous atoms and any other spot "between these centres? a difference in degree, or even in the nature of the power consistent with the law of continuity, I can admit, but the difference between a supposed little hard particle and the powers around it I cannot imagine.

To my mind, therefore, the *a* or nucleus vanishes, and the substance consists of the powers or *m*; and indeed what notion can we form of the nucleus independent of its powers? all our perception and knowledge of the atom, and even our fancy, is limited to ideas of its powers: what thought remains on which to hang the imagination of an *a* independent of the acknowledged forces? A mind just entering on the subject may consider it difficult to think of the powers of matter independent of a separate something to be called *the matter,* but it is certainly far more difficult, and indeed impossible, to think of or imagine that *matter* independent of the powers. Now the powers we know and recognize in every phenomenon of the creation, the abstract matter in none; why then assume the existence of that of which we are ignorant, which we cannot conceive, and for which there is no philosophical necessity?

Before concluding these speculations I will refer to a few of the important differences between the assumption of atoms consisting merely of centres of force, like those of Boscovich, and that other assumption of molecules of something specially material, having powers attached in and around them.

With the latter atoms a mass of matter consists of atoms and intervening space, with the former atoms matter is everywhere present, and there is no intervening space unoccupied by it. In gases the atoms touch each other just as truly as in solids. In this respect the atoms of water touch each other whether that substance be in the form of ice, water or steam; no mere intervening space is present. Doubtless the centres of force vary in their distance one from another, but that which is truly the matter of one atom touches the matter of its neighbours.

Hence matter will be *continuous* throughout, and in considering a mass of it we have not to suppose a distinction between its atoms and any intervening space. The powers around the centres give these centres the properties of atoms of matter; and these powers again, when many centres by their conjoint forces are grouped into a mass, give to every part of that mass the properties of matter. In such a view all the contradiction resulting from the consideration of electric insulation and conduction disappears.

The atoms may be conceived of as highly *elastic,* instead of being supposed excessively hard and unalterable in form; the mere compression of a bladder of air between the hands can alter their size a little; and the experiments of Cagniard de la Tour carry on this change in size until the difference in bulk at one time and another may be made several hundred times. Such is also the case when a solid or a fluid body is converted into vapour.

With regard also to the *shape* of the atoms, and, according to the ordinary assumption, its definite and unalterable character, another view must now be taken of it. An atom by itself might be conceived of as spherical, or spheroidal, or where many were touching in all directions, the form might be thought of, as a dodecahedron, for any one would be surrounded by and bear against twelve others, on different sides. But if an atom be conceived to be a centre of power, that which is ordinarily referred to under the term *shape* would now be referred to the disposition and relative intensity of the forces. The power arranged in and around a centre might be uniform in arrangement and intensity in every direction outwards from that centre, and then a section of equal intensity of force through the radii would be a sphere; or the law of decrease of force from the centre outwards might vary in different directions, and then the section of equal intensity might be an oblate or oblong spheroid, or have other forms; or the forces might be disposed so as to make the atom polar; or they might circulate around it equatorially or otherwise, after the manner of imagined magnetic atoms. In fact nothing can be supposed of the disposition of forces in or about a solid nucleus of matter, which cannot be equally conceived with respect to a centre.

In the view of matter now sustained as the lesser assumption, matter and the atoms of matter would be mutually penetrable. As regards the mutual penetrability of matter, one would think that the facts respecting potassium and its compounds, already described, would be enough to prove that point to a mind which accepts a fact for a fact, and is not obstructed in its judgement by preconceived notions. With respect to the mutual penetrability of the atoms, it seems to me to present in many points of view a more beautiful, yet equally probable and philosophic idea of the constitution of bodies than the other hypotheses, especially in the case of chemical combination. If we suppose an atom of oxygen and an atom of potassium about to combine and produce potash, the hypothesis of solid unchangeable impenetrable atoms places these two particles side by side in a position easily, because mechanically, imagined, and not unfrequently represented; but if these two atoms be centres of power they will mutually penetrate to the very centres, thus forming one atom or molecule with powers, either uniformly around it or arranged as the resultant of the powers of the two constituent atoms; and the manner in which two or many centres of force may in this way combine, and afterwards, under the dominion of stronger forces, separate again, may in some degree be illustrated by the beautiful case of the conjunction of two sea waves of different velocities into one, their perfect union for a time, and final separation into the constituent waves, considered, I think, at the meeting of the British Association at Liverpool. It does not of course follow, from this view, that the centres shall always coincide; that will depend upon the relative disposition of the powers of each atom.

The view now stated of the constitution of matter would seem to involve necessarily the conclusion that matter fills all space, or, at least, all space to which gravitation extends (including the sun and its system); for gravitation is a property of matter dependent on a certain force, and it is this force which constitutes the matter. In that view matter is not merely mutually penetrable, but each atom extends, so to say, throughout the whole of the solar system, yet always retaining its own centre of force. This, at first sight, seems to fall in very harmoniously with Mossotti's mathematical investigations and reference of the phenomena of electricity, cohesion, gravitation, &c. to one force in matter; and also again with the old adage, "matter cannot act where it is not." But it is no part of my intention to enter into such considerations as these, or what the bearings of this hypothesis would be on the theory of light and the supposed æther. My desire has been rather to bring certain facts from electrical conduction and chemical combination to bear strongly upon our views regarding the nature of atoms and matter, and so to assist in distinguishing in natural philosophy our real knowledge, *i. e.* the knowledge of facts and laws, from that, which, though it has the form of knowledge, may, from its including so much that is mere assumption, be the very reverse.

I am, my dear Sir, yours, &c.,

MICHAEL FARADAY.

Source Endnote

1 Lond. and Edinb. Phil. Mag., 1844, vol. xxiv. p. 136.

Editorial Headnote

20. Charles Wheatstone letter to Alexander Bain, June 13, 1842, in *An Account of Some Remarkable Applications of Electric Fluid to the Useful Arts* (London: Chapman and Hall, 1843), pp. 116–121.

This series of letters from Alexander Bain (1810–1877) to Charles Wheatstone (1802–1875) appeared in a volume by John Finlaison (1783–1860) that aimed to name Bain as the *first* inventor of the electro-magnetic printing telegraph. Bain apprenticed as a clockmaker in Edinburgh and later with John Barwise Jr. of the famed Barwise Watch and Clockmakers. He is most famous for his work on telegraphy and the patent of the electric clock. Bain also engaged in a feud with Wheatstone over the patent of the printing telegraph. Finlaison's volume contains Bain's writing and illustrations explaining about his inventions and the potential uses of electricity, and also a detailed timeline of Bain's patent feuds. In the appendix, Finlaison included several of Bain's letters. These letters also appeared in the *Literary Gazette*, a British literary magazine that featured book reviews, social columns, theatre announcements, and "original correspondences." The first letter shows Bain's claim that he authored the invention of the Electro-Magnetic Printing Telegraph. He challenges Wheatstone to answer his claim and says he's done so before in the *Inventors' Advocate,* a quarterly targeted at parties interested in the mechanical arts. This is followed by a retort by Wheatstone in which he explains his encounters with Bain and his own inventions. The series of exchanges continues and includes both men's correspondences with others supporting their side of the story. Bain uses his expertise and presentations he made in the *Mechanics Magazine,* while Wheatstone relies on his presentation to the Royal Society and close correspondences with others engaged in electromagnetic theory. As Finlaison mentions earlier in the book, "it is certainly possible that one man may have a *legal* right to an invention, while the *moral* claim belongs to another." Although Bain's patent claims succeeded, he was unable to profit off of them. However, the resolution of this controversy is secondary to the discourse showcased in these letters. They highlight the breadth of "scientific gentlemen," mechanics, artisans, and public audiences privy to claims of invention. Furthermore, this discussion showcases the epistemic tension surrounding electromagnetic theory and electric apparatus. Wheatstone sought to diminish Bain's knowledge, while Bain worked to point out Wheatstone's reliance on others with mechanical knowledge.

20

CHARLES WHEATSTONE LETTER TO ALEXANDER BAIN, JUNE 13, 1842

Source: Charles Wheatstone letter to Alexander Bain, June 13, 1842, in *An Account of Some Remarkable Applications of Electric Fluid to the Useful Arts* (London: Chapman and Hall, 1843), pp. 116–121.

NO. 2

From the LITERARY GAZETTE, 18th June, 1842.

"SIR,

"IN answer to a letter, signed "Alexander Bain," which appeared in the *Literary Gazette* of Saturday last, I beg to offer the following observations:

"The writer states that the electro-magnetic telegraph clock is not my invention, and that I have not dared to claim it openly. To this my answer is, that on November the 26th, 1840, a paper of mine was read at the Royal Society, fully describing it as my invention; and the telegraph-clock itself was shown in action in the library on the same evening, and during several days. An abstract of this paper was published in the society's proceedings; and the communication was noticed in the *Literary Gazette* of 28th November, 1840. It was not until the January following that I became aware that an attempt was about to be made to question my right to the invention, by receiving a notice from a Mr. Barwise, of St. Martin's Lane, stating that he was the inventor: some time after which it was publicly announced in the placards and advertisements of the Polytechnic Exhibition as being the joint invention of Messrs. Bar-wise and Bain. The latter person was a working mechanic, who had been employed by me between the months of August and December of the year 1840.

"It admits of no doubt, therefore, that this invention was first publicly made known and claimed by myself; and I proceed to the assertion of the writer, that he communicated the invention to me in August, 1840, which was three months preceding the date of my publication. To this I answer, that there is no essential difference between my telegraph-clock and one of the forms of the electro-magnetic telegraph, invented by me, and described in the specification of a patent granted to myself and Mr. Cooke in January, 1840; the former is one of the numerous and obvious applications which I have made, and only requires the idea of telegraphing time to present itself for any workman of ordinary skill to put it into practice. In telegraphing messages, the wheel for making and breaking the circuit is turned

DOI: 10.4324/9781003295556-23

round by the finger of the operator, while in telegraphing time it is carried round by the arbor of a clock. The sole question, then, is, Did the idea of applying my invention to telegraph time originate with myself, or was it suggested to me by your correspondent? Now, with reference to this, I have to state, that long before the date specified I had described to many of my friends in what manner the principles of my telegraph might be applied, to enable the time of a single clock to be shown simultaneously in all the rooms of a house, or in all the houses of a town. Among these the following gentlemen have, from particular circumstances, been able to furnish me with the dates of the communications I made to them: Mr. Airy, the astronomer royal; Dr. W. A. Miller, of King's College; Mr. John Martin, the eminent artist; and Mr. F. O. Ward, formerly a student in King's College. In addition to this evidence, I may add, that Mr Bain's letter, in the *Inventors' Advocate,* was immediately answered by Mr. Lamb, a workman in my employ, to the purport, that it was impossible the statement therein contained could be true, since I had given him instructions to make the electro-magnetic telegraph-clock on January 6, 1840, which was more than six months before Mr. Bain asserted he made his communication. I repeat that, neither as regards the idea, nor any of the details of the telegraph-clock, have I been in the slightest degree indebted to your correspondent; and I think, Mr. Editor, you must allow that I have satisfactorily refuted his assertions.

"I next proceed to the consideration of my electro-magnetic printing-telegraph. This invention consists merely of an addition to the electromagnetic telegraph invented by me, and described in the first part of the specification of the patent granted to myself and Mr. Cooke in January, 1840; when this addition is removed, the telegraph itself remains, in all its details, without the slightest alteration. There cannot, therefore, be a question as to the invention of my printing telegraph as a whole, but merely as to the additional apparatus which occasions the letters to be printed, instead of their being merely presented to the eye. The following are the means by which I effect this purpose: – For the paper disc of the telegraph, on the circumference of which the letters are printed, a thin disc of brass is substituted, cut from the circumference to the centre, so as to form four-and-twenty springs, on the extremities of which types or punches are fixed; this type-wheel is brought to any desired position, just as the paper disc is. The additional part consists of a mechanism which, acted upon by an electro-magnet, occasions a hammer to strike the punch, brought opposite to it, against a cylinder, round which are rolled alternately several sheets of thin white paper and of the blackened paper used in the manifold-writing apparatus; by this means, without presenting any resistance to the type-wheel, I obtain at the same time several distinct printed copies of the message transmitted. This plan originated with and has been carried out solely by myself. It is true, that after I had contrived this arrangement, Mr. Bain proposed a different and far less efficient mode of effecting the same purpose: this was, to move the type-wheel bodily towards the cylinder, round which wet paper was to be rolled to receive the impression, and to employ an inking-apparatus similar to that of the common printing-machine, in order to supply the type-wheel with ink. Though I purchased from him the rude model which he made to explain his notions, and subsequently employed him to see how far it was capable of practical application to my telegraph, I have never made, nor do I ever intend to make use of any of his suggestions, nor have I ever laid the slightest claim to them. The truth of this statement admits of no dispute, since, in a receipt in my possession, signed and dated by himself on August 18, 1840, he acknowledges receiving payment for a model of his proposed "modification of the printing-apparatus to be added to the electric telegraph," and admits that 'it differs from the instrument devised by myself for the same purpose by an inking-roller being employed, and by the wheel, on the circumference of which the types are placed, being bodily moved forward, in order to impress the types on the cylinder carrying the paper, instead of the types being pressed individually, as in my instrument' Whatever may be the merits of Mr. Bain's method, it cannot justify any person to call into question the originality and priority of my electro-magnetic printing-telegraph, which is secured from infringement by two patents, one, already mentioned, including the telegraph itself; and the other, of more recent date, comprising my superadded printing-apparatus.

"As I have advanced nothing in this communication but what can be supported by documents, or the evidence of other persons, I shall consider any further unsupported assertions unworthy of notice, and shall therefore decline any further correspondence.

"I remain your obedient Servant,

"C. WHEATSTONE."

"*King's College, June* 13,1842."

NO. 3

From the LITERARY GAZETTE, 6th August, 1842.

"SIR

"IN order to counteract the prejudicial tendency of Professor Wheatstone's letter in your *Gazette* of the 18th of June, as well as to enable your readers to form a correct judgment upon the points at issue between the Professor and myself, I have to request a place for the following brief but faithful narrative. I first came to London in 1837, to seek employment as a journeyman clockmaker. I had at that time some knowledge of electricity, and a strong desire to know more. As soon as I obtained employment, I devoted all my leisure to my favourite study, and attended lectures at the Adelaide Gallery and Polytechnic Institution; and seeing the beautiful electro-magnetic machines in action at those places, first drew my attention to how they could be applied to useful purposes. The application of this mysterious power to the mechanism of my own business was naturally the first to suggest itself; afterwards I thought on various ways of applying it to telegraphs; and, among other methods which occurred to me, was that of printing the intelligence, instead of showing it by signs, which I knew had been done before. I therefore confined my exertions to the electric clocks and electric printing-telegraph; and by July, 1840, I had so far matured both these inventions, that I was desirous of meeting with some party who would assist me with the means of bringing them into operation. Being a stranger in London, I called, first, on Sir Peter Laurie (as a countryman of mine), who kindly gave me a letter of introduction to the late Dr. Birkbeck; the Doctor was unfortunately confined to a sick-bed, and I never saw him. I next called at the *Mechanics' Magazine* office, in Fleet-street, and saw Mr. Baddeley (a gentleman well known in the scientific world), to whom I explained my wants. Mr. Baddeley advised me to call upon Professor Wheatstone, whom he knew to be deeply engaged in the science of electro-magnetism, and thought him a likely person to enter into my views. At that time I knew nothing of Professor Wheatstone; but at Mr. Baddeley's recommendation I waited upon him and described my plans, in which he seemed to take great interest.

"At a second interview, I exhibited a model illustrative of my printing-telegraph, and another of my electric clock: from the remarks then made by Professor Wheatstone, it was evident that both inventions were entirely new to him, nor did he in any way question their novelty or originality. It appeared subsequently, however, that as soon as he got possession of my plans, he went to another workman, and got a machine made, which he exhibited at the Royal Society as his own invention, well knowing that he had obtained it from me: but of these proceedings I was kept in ignorance.

"The model illustrative of my printing-telegraph, which I exhibited to Professor Wheatstone at our second meeting, he purchased of me; but he said that he also had a printing-machine which was to be added to his telegraph; and under this impression I sold him my printing model, and signed the receipt mentioned in his letter, the receipt being written by himself. At that time I had every confidence in the Professor, and implicitly believed his statements. Shortly afterwards I showed him another printing-telegraph, different from the first, when we came to the following arrangement, viz. – That I was to make working models of my printing-telegraph (including the two arrangements), for which I was to be remunerated as the work progressed: the invention was to become his property, on his paying me a stipulated sum of money. Accordingly I proceeded with the work, until one of the models was finished, and the other partly so; when the Professor contrived to get both into his possession, and then refused to fulfil his part of the agreement . . . .

"Shortly afterwards, I learnt that Professor Wheatstone was applying for a new patent, the title of which would have enabled him to appropriate to himself my inventions; to prevent this, I immediately made fresh models, and publicly exhibited them at the Polytechnic Institution.

"Professor Wheatstone's assertion, that I 'was employed by him as a working mechanic,' is wholly untrue; the only transactions which ever took place between us related entirely to my own inventions, as I have already described; and I was never engaged or employed for a single hour upon any piece of mechanism or invention emanating from Professor Wheatstone.

"The Professor asserts, that there is no essential difference between some part of his patent electric telegraph of 1840, and the electric clock – that it only requires the idea of telegraphing time instead of signals; forgetting that in the idea and its application consists the fundamental part of the electric clock. The very principle of the invention consists in applying the synchronism of the pendulum to regulate the transmission of the electric currents. Professor Wheatstone states, that he directed his workman, in January 1840, to make an electro-magnetic telegraph clock; but it was *not* made, nor anything done in the matter, until after I had communicated my method to him; it was then gone on with so quickly that several were soon completed.

"Professor Wheatstone states, that he described to many of his friends how the principle of his telegraph could be applied to clocks, so as that the time of a single clock should be shown simultaneously in all the houses of a town. This assertion proves him to be a professor of a practical impossibility, and shows that he does not even yet understand much of the invention. To work the clocks of a very small town, simultaneously, would require a source of electricity of such immense power, that instead of being able to transmit it through wires, it would make bars of metal hot in a few minutes. In the electric clocks of Mr. Barwise and myself this difficulty is got over by working the clocks in rotation.

"Professor Wheatstone took out a patent in January 1840, in which he might have secured the inventions of the printing-telegraph and the electric clock, had he been acquainted with either previous to lodging his specification in July, 1840. Any body who knows the Professor will agree with me that, had he known of these inventions, he would have secured them in his patent, the more especially as they are far more valuable than what he patented.

"The specification of that patent is open to public inspection, and does not contain the slightest allusion to either, which goes far to prove that his assertion of having a printing machine when I showed him mine (in August) must have been entirely destitute of truth. The only features of resemblance between the patented electric telegraph of Mr. Wheatstone and my electric clock are, that they both contain electro-magnets, and wheels and pinions – and so did the electric telegraph of Mr. Davy, patented in 1838.

"Did your space permit, I could state many other circumstances connected with the inventions of which Professor Wheatstone has endeavoured to deprive me; but, in conclusion, will merely request insertion of the two following letters, which are corroborative of the first and most important part of my narrative. – I remain, Sir, your obedient and obliged humble servant,

"ALEXANDER BAIN."[1]

"320, *Oxford Street, June* 29, 1842.

Source Endnote

1 We regret having got entangled in this controversy, and shall only hold ourselves in justice bound to afford a place to any reply Mr. Wheatstone may think proper to give. – *Ed. L. G.*

Editorial Headnote

21. "Industrial Applications," in *Electric Lighting: A Practical Treatise*, Paget Higgs, trans., (London: E. & F. N. Spon., 1878), 103–125.

In 1873, the Belgian engineer Zénobe Gramme (1826–1901) and French engineer Hippolyte Fontaine (1833–1910) displayed the Gramme Dynamo, sometimes known as the Gramme Machine at the Vienna Exhibition. The direct current dynamo proved to be the first electric motor capable of powering machinery for industrial purposes. Both Fontaine and Gramme demonstrated the efficacy of their work throughout Europe demonstrating the uses of such a motor and long-distance power transmission. During the 1870s and 1880s, numerous texts advocated for these new applications. Here is a volume originally published in French and translated by the British electrical engineer Paget Higgs. The book comments on current electromagnetic machinery, lighting, and details the uses of the Gramme Machine. In the selected chapter, Fontaine focuses on the industrial potential of electrically powered machinery and factories. For instance, he celebrates the economic and visual efficiency of the electric light for companies that need illuminated workspaces and cites Gramme's factory and the Ducommun foundry as models. In another factory mentioned in this section, the Ménier Caoutchouc (Un-vulcanized rubber) Factory in Grenelle, France, the proprietor, Émile-Justin Menier (1826–1881), claims that three electric lamps replaced 259 gas-burners, foreshadowing the brewing competition between gas and electric lighting in coming decades. Beyond factory floors, these pages highlight the utility of electric lighting for spinning mills, ports, and canals. By the late 1870s, in the minds of boosters, electric lighting needed serve industrial and profitable ventures. The light was no longer relegated to the wealthy's parlors or entertainment halls. This work and the many, many other tracts concerning lighting at the time denote a major shift in how people consumed information about the electric light, and most especially how scientists and inventors discussed the applications of electricity.

21

INDUSTRIAL APPLICATIONS

Hippolyte Fontaine

Source: Hippolyte Fontaine, "Industrial Applications," in *Electric Lighting: A Practical Treatise*, Paget Higgs, trans., (London: E. & F. N. Spon., 1878), 103–125.

LIGHTING by electricity may be advantageously employed in a large number of works. It offers no practical inconvenience and admits of obtaining a great quantity of light at small expense. It is the only industrial illumination with which there can be executed by night as by day, the loading and discharging of cargoes, the mounting of machinery, carpentry, weaving, &c., and easy superintendence in a workshop attained. The light produced is in such abundance that, reflected from all the objects it falls upon, it is diffused in all directions like daylight; there is no part absolutely dark; it can be read by, and tools found by it, &c.

In spite of the luminous power of a focus, if work of a certain accuracy is to be done by it, it is indispensable to have two machines, so that the shadows produced by one light may be illuminated by the other; but then the extinction of light in each lamp necessary at the end of three-and-a-half to four hours' illumination is not a serious inconvenience; two minutes are sufficient to replenish the carbons in the lamp and to relight them, and during this short interval of time, the light from the other lamp prevents interruption of the work. Besides, if it is indispensable to have absolute continuity, there should be added to each lamp another for interchange, the lighting of the second lamp occurring automatically by the extinction of the first; but in practice this is not generally necessary.

Experience has shown that working by the naked electric light is not fatiguing to the eyes; after for some days using opal globes to temper the light, these were removed at the request of the workmen. The reduction of the light is unnecessary.

It is generally known that the electric light preserves the tints of colours. This property has been utilised with success by several dyers for standardising their colours by night; a single luminous focus of the very smallest dimensions is sufficient in this case.

When the ceilings are at less height than four mètres, the introduction of the electric light becomes more difficult, without, however, being absolutely impossible.

Generally, there may be conveniently lighted with a single apparatus, 500 square mètres of fitters' shops, lathe-shops, tool-shops, modelling-rooms, &c.; 250 square mètres in a spinning-mill, weaving establishment, printing-rooms, &c., and 2000 square mètres of yard, court-yard, dockyard, quay, and open-air works, &c. With these data, it is easy to calculate the cost of introduction anywhere, knowing that a complete apparatus, lamp, machine, conducting wire, transport and mounting, costs about 2400 francs (= 100*l.*) in France and its frontier countries.

Manufacturers who pay only 0·30 franc per cubic mètre, and find their establishments sufficiently lighted with twenty gas burners, should not seek a more economical light, unless they work all night without interruption; in which case it would be to their interest to replace ten gas burners by an electric apparatus.

The two questions to be examined are the cost price and the convenience. The second is often neglected, but this is wrong; because in the majority of cases, it is of the greater importance. We have had occasion to visit a large number of works, factories, and manufactories of all kinds in Europe and in America, and we have seen but few well-lighted establishments; few where a single more intense light would not have given easier superintendence, greater amount of work, and greater security. Fire insurance companies have so important an interest in the introduction of the electric light, that several have offered to lower their tariff for all buildings thus lighted.

A rapid review of some introductions already made will lead to better appreciation than any possible reasoning as to the advantages of electricity.

The Gramme Company's Workshops

The first permanent introduction of the electric light was realised in 1873, at Paris, in the workshops of the Gramme Company. The light was furnished from a single focus, which took the place of twenty-five gas burners. For four years its use has been regular and the cost price has not exceeded 0·60 franc per hour, including all expenses. The space lighted is about 12 mètres by 12 mètres and 5 mètres in height.

The Ducommun Establishment at Mulhouse

In consequence of a visit to the Gramme workshops, Messrs. Heilmann and Steinlen, the present proprietors of the Ducommun factories at Mulhouse, decided to apply this new method of lighting upon a larger scale. For this purpose they placed in their iron foundries four Serrin lamps fed by four Gramme machines. The trial was crowned with success; during three years the working of the apparatus has been regular, and Messrs. Heilmann and Steinlen have the intention of extending this system of lighting to several other buildings.

The Ducommun foundry is of recent construction. It is well arranged for electric lighting without in any way being specially intended for this purpose. (The construction was entirely finished when it was decided to introduce the electric light.) It is a large hall without partition walls or vertical screens, of 56 mètres interior length and 28 mètres width. Two large travellers for the manipulation of castings and crucibles circulate from one extremity of the building to the other. At

5·5 mètres from the floor, level with the travellers, running along the two longitudinal sides, is a platform of some mètres width. The roof is a double incline, the joinery of which is slight, and the walls are lime-washed.

The Serrin regulators are placed on two light stands jutting out laterally from the platforms. Access is gained to them by a ladder, as with public lamps in gas lighting. The foci are raised 5 mètres above the floor; they are 21 mètres apart in the direction of the length of the building, and 14 mètres in the direction of the width.

The Gramme machines are placed in the engine-house with the machine employed for ventilation. The motor being larger than required for the work, alteration was unnecessary, and the ventilator and the electric apparatus are actuated simultaneously without inconvenience. The introduction has been very simple. The Gramme machines are placed on the first floor in one line, and they receive motion from a single intermediate shafting. All the parts are easily accessible, and the maintenance, which is reduced to some precautions in cleaning, can be effected during working, as well as when the machines are at rest. The motor is of the Sulzer type, its working is very regular, and the consumption of fuel small.

The lighting is generally constant in intensity; at any point of the place it is easy to read writing at normal distance from the eyes. There are nearly no shadows, because of the cross rays from the four lamps.

The complete introduction cost, in round numbers, 10,000 francs (= 400*l.*); which is about the cost of introducing 250 gas burners. The total light produced by electricity exceeds 400 burners.

We presently give the results of experiments made by Messrs. Schneider and Heilmann on the motive power absorbed by the machines, the lighting power of each lamp, and the cost of electric lighting as compared with gas lighting.

Messrs. Heilmann and Steinlen have obtained from the Gramme Company the right of manufacturing machines of a given type, and they have studied several arrangements to facilitate the use on ships and in war manœuvres. Fig. 36 shows one of these arrangements. It is a Gramme machine placed on the same sole as the steam motor and intended for introduction into sugar-refineries, on board ships, especially where steam is furnished by ordinary boilers, and where the motor of the electric apparatus may be independent to ensure regularity.

To avoid all vibration, the sole is large, rigid, and well set. The total volume does not exceed 2·25 cubic mètres on 1 mètre. All friction bearings are calculated for durable service. The shafting of the steam-motor makes 150 revolutions per minute; that of the Gramme machine 850.

A rider, placed under the electric machine, admits of drawing it farther from the motor when the belt has not sufficient tension. This is an excellent means of preventing slipping of the belt, and consequently of ensuring regular velocity of the coil, but its use needs considerable care; an exaggerated tension of the belt absorbs much work and occasions greater inconvenience than the slipping itself.

The cost of the complete apparatus, motor and electric generator, is 4000 francs (= 160*l.*).

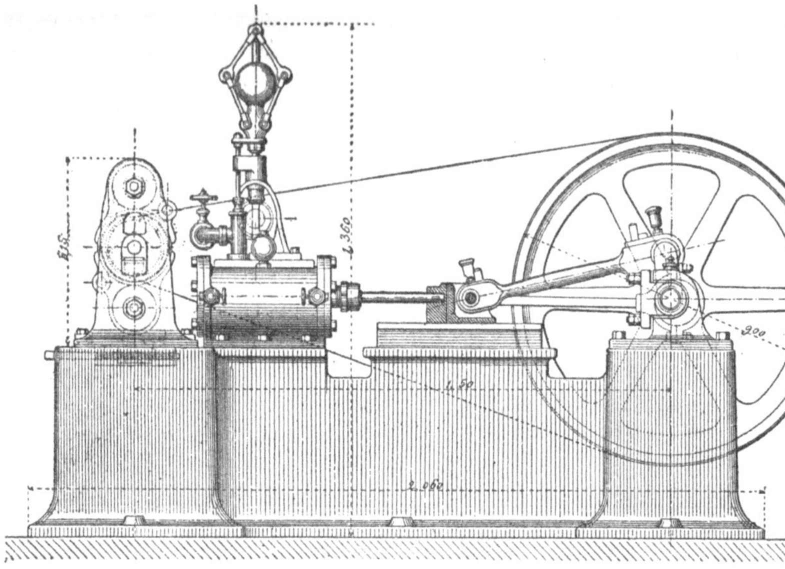

Fig. 36. Gramme Machine with its Motor.

Sautter, Lemonnier, and Co.'s Workshops at Paris

Electric lighting is an important feature in the workshops of Messrs. Sautter, Lemonnier, and Co., the well-known makers of lighthouse-lenses, because no one than the members of this firm is more familiar with the use of strong electric lights, and, from the time of M. Gramme's first attempts, they asked and obtained authority to construct machines upon his system.

Its introduction into their workshops admitted that they could convince themselves and demonstrate to visitors that this method of lighting was excellent from all points of view, and that under certain conditions of local arrangements and for certain kinds of work, it is to be preferred to all others.

The Sautter and Lemonnier workshops consist of two buildings, each 30 mètres in length by 25 mètres in width; a platform, at 6 mètres from the flooring, runs between these buildings, with a width of 10 mètres. On the ground-floor are the machine tools, lathes, planing machines, drills, punching machines, shears, forges, &c.; the fitters there operate upon the larger work, and mount machines, &c. On the first floor are the model-makers, tinmen, and bronze-work fitters, with finer apparatus; three Gramme machines, giving each a light equivalent to 150 Carcel burners, illuminate the shops so effectually that all other methods of lighting have been removed, as well for operators with machine tools as for those doing finer work.

It might easily be imagined that in a shop of this kind the shadows thrown by the lathes, belts, and columns would be of great inconvenience to the work, and

Fig. 37. MESSRS. SAUTTER, LEMONNIER, AND CO.'S WORKSHOPS.

that especially where the light does not fall directly, the contrast would make the obscurity appear greater than with another method of lighting. This is not so. The light diffused, or, in other words, the light reflected by all points illuminated is such that there is not, properly speaking, at any part of the workshop a dark corner, and that the workman may easily distinguish objects at the bottom of his drawer.

The electric light, no more than any other, neither injures nor fatigues the sight of the workmen; they lose in a few days the habit of looking at it, and are very satisfied with a method of lighting that renders nightwork as easy as work by day.

A still stronger reason may be given with regard to superintendence, which can be exercised by night in a workshop lighted by the electric light absolutely as well as by day.

Intermittences are not to be feared, and this is the case in every workshop lighted by at least two lamps. When these lamps are well regulated, accidental extinctions are very rare. At the end of four hours' illumination, it is necessary to replace the carbons, and this operation is performed in two or three minutes.

Messrs. Sautter, Lemonnier, and Co. have found that each Gramme machine consumes work equal to about 2-HP; this and the carbons for the lamp form the only expenditure for lighting. The carbons are consumed at the rate of 0·07 mètre per hour, and cost 2 francs per mètre. 100 Carcel burners produced by this machine cost 0·14 franc per hour, in addition to the cost of motive power.

Fig. 37 is a view of the Sautter and Lemonnier workshops lighted by electricity. The engraving shows the whiteness of the ceiling and the uniform intensity of the lighting. It gives as exact an idea as possible of the remarkable effects of the voltaic arc in a closed and covered space.

Ménier Works

M. Ménier has introduced the electric light in his various establishments, and has made good use of it in several particular arrangements. Since November, 1875, 14 machines, of 150 burners, have been in use, entirely to his satisfaction: 3 are set up in the sugar refinery at Roye, 3 in the caoutchouc factory at Grenelle (Paris), and 8 in the celebrated Ménier Chocolate factory at Noisiel.

In all these works, the lamps are so suspended as to admit of the renewal of carbons when necessary, without use of ladder or steps.

The roller represented in Fig. 38 is the invention of M. Henri Ménier. It consists of two cast-iron cheeks mounted on a hardwood slab, and of a drum of vulcanite. The conducting wires are attached one to the left cheek, the other to the right cheek, and these communicate metallically with the extremities of the suspending cable.

This cable is formed (Fig. 39) of an exterior covering of hemp, of a sheathing of caoutchouc, a series of copper wires braided like a wick, of a second sheathing of caoutchouc, and finally of a series of copper wires stranded into a cord, and forming the core of the cable. The winding on the drum is easily effected. A small ratchet prevents the lamp descending by itself.

The cable, sustained and guided by two upper pulleys, is attached to a small plate which is connected to the lamp by two curved bars (Fig. 40). A frame, mounted with two lugs, is fixed to the lamp, and receives the extremities of the curved bars.

The current is led to the terminals of the lamp by the suspension bars, one of which is connected to the metallic core of the cable, and the other with its intermediate metallic part, that is to say, with the positive and negative poles of the magneto-electric machine.

The Noisiel works are lighted all night. The eight machines are coupled in two batteries of four. They are put in motion by hydraulic wheels, and, if necessary, when the waters are low, by a special steam motor. The wires are led to a commutator placed on the ground-floor (in the centre of several workshops), which admits of sending the current from each machine in fifteen different directions. By this means a room can be lighted by any machine, and the inconvenience arising from one of the machines or lamps getting out of order is entirely avoided.-

Near each lamp is an interrupter, with which the lamp may be extinguished without stopping the motion of the machines.

At Noisiel the luminous foci are thus distributed:

> A lamp placed in a square lantern and elevated 7 mètres above the level of the steam motor, lights a principal yard of 2000 square mètres surface.
> Two other lamps each light an interior yard of 500 square mètres.

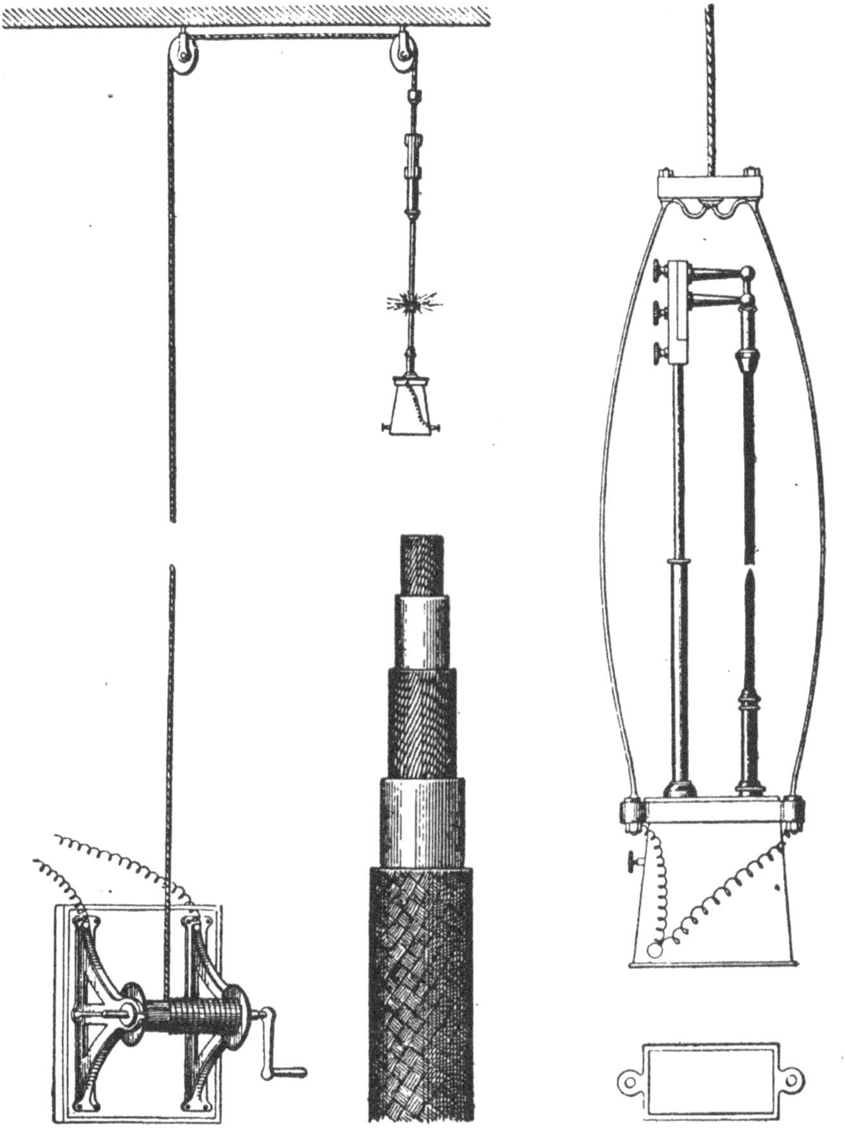

Fig. 38. *Fig. 39.* *Fig. 40.*

MÉNIER'S SYSTEM OF LAMP SUSPENSION.

The torrefaction shop, where 32 operators work, is 44 mètres long, 11 mètres wide, and 7·70 mètres high; it is lighted by a single lamp placed in a glazed lantern on the ground, at one of the extremities of the workshop. The light is projected on to the ceiling by a parabolic mirror, inclined somewhat, and is thus diffused throughout by reflection. The weighing and moulding shop, where are 90 workmen, is 52 mètres in length by 11 mètres in width, and 7·70 mètres in height. It is lighted by two lamps placed at 25 mètres distance from each other, and suspended at 6 mètres from the ground.

The mechanics' shops are of 400 square mètres surface; these are lighted by a single lamp, suspended 6 mètres from the ground.

The grinding shops, which are 40 mètres long, 12 mètres wide, and only 3·60 mètres high, are lighted by reflection.

M. Ménier possesses, undoubtedly, the best system of electric lighting that exists in the world. He can, without any derangement, lead this light into his residence and light the hall, the large dining-room, the two drawing-rooms, the winter garden, and the gardens.

One lamp is specially intended for photometric projections and for scientific demonstrations in the lecture and conference hall.

On the occasion of the first distribution of prizes from the School of Industry, M. Ménier gave, 20th August, 1876, a banquet of 1450 covers in a tent 65 mètres long, 32·50 mètres wide, and 7 mètres high. The banquet and the ball that followed were lighted in a fairy-like manner by six regulators of 150 burners. A lighthouse of 1200 burners, placed at some distance from the tent, illuminated the country for a radius of more than a kilomètre.

At the central sugar-refinery at Roye, M. Ménier has set up three Gramme machines with similar arrangements to those adopted at Noisiel. The first of these machines lighted the "carbonatation" room, which is 45 mètres long, 20 mètres wide, and 15 mètres high; the second is placed in the middle of the principal yard, where are the lime-kilns, stables, coal-sheds, &c.; the third lights the beet-root receiving depôt and washing-rooms.

At Grenelle, the caoutchouc factory has its large hall, where 150 persons work, lighted by three regulators suspended in a triangle at 6 mètres from the ground. The hall is 47 mètres long by 41 mètres wide. The machines are worked sometimes by the general motor, sometimes by a small special machine.

The three electric lamps replace 259 gas-burners, and produce, according to M. Ménier, five times more light than gas gives. At Grenelle, as at Noisiel and at Roye, the same system of raising the lamp by a cable with two conductors and the same arrangement of commutator is employed.

Spinning Mills

Among the introductions of the electric light into spinning-mills may be mentioned those of Madame Dieu-Obay, at Daours; Messrs. Ricard and Sons, at Manresa, Barcelona; and those of Messrs. Buxeda Brothers, at Sabadell, Spain.

1. *Madame Dieu-Obay's Mills.* – The workshop is 3·70 mètres high, 43 mètres long, and 11 mètres wide. It includes nine double looms, with thirty needles, and seventeen spinning-mills, with fifty-two jockeys. Fifty persons, workmen and women, are employed.

The light is obtained by two Gramme machines worked by a hydraulic motor, which also actuates the looms.

The lamps are suspended at a height of about 2 mètres, and these are each furnished with a large lamp-shade reversed. This shade projects the luminous rays up to the ceiling, and these rays are thus dispersed in all directions. The focus is not visible, and consequently the sight rests on strongly illuminated, but not on incandescent, surfaces.

It has already been stated that the same artifice has been employed with success at M. Ménier's at Noisiel. The only special arrangement at Daours is in the height of the lamps, which is nil at Noisiel and mid-height at Daours. The result is very good in both cases.

Wherever there is a very white ceiling, and at a height of at least 4 mètres, there is an advantage in employing reflected light instead of the direct light.

2. *Messrs. Ricard and Son's Cotton Mills.* – The first floor of these mills is 33 mètres long by 21·20 mètres wide. Two lamps light ten self-acting machines. The Gramme machines are placed at the end of the workshop and are actuated by the general motor with an intermediate shafting. The height of the lamps is 3·40 mètres and their relative distance 15 mètres. We represent this arrangement (Fig. 41) because it is made under the very unfavourable condition of lowness of ceiling; it is the possible limit for good results, which it is requisite to exceed to preserve to the electric light all its advantages.

The second floor, the part occupied by the looms, is 16 mètres long by 21·20 mètres wide. A single lamp suffices. There are here five self-acting mules working all night, like those of the first floor, since May, 1876. The proprietors are very satisfied with their lighting.

3. *Linen Looms and Mills* (*Messrs. Buxeda Brothers*). – The room lighted electrically at Messrs. Buxeda's works is 58 mètres long by 22 mètres wide. It contains thirteen spinning-machines, twelve carding-machines, one loom, and some accessory machines. The lamps, three in number, are suspended at a height of 4·20 mètres, and distant from each other 13 mètres. There are four workmen to each spinning-machine, three to each group of two carding-machines, and ten to the accessory apparatus; in total, eighty workmen in the same room.

With gas lighting certain work could not be done at night, because of the differences observed in the colours; now work is carried on night and day without stopping and without the least inconvenience in spite of the colour-tints.

The introduction at Messrs. Buxeda's was made six months ago (1877). The better quality of the products fully compensate the expenditure of motive power. The lighting itself is much more economical than that with gas or petroleum.

Chapelle-Paris Goods Depôt

In consequence of experiments made in 1876 at the railway station in Paris, the Northern Railway Company decided upon the application of the electric light in the goods depôt, where work is going on all night. As it was necessary to establish

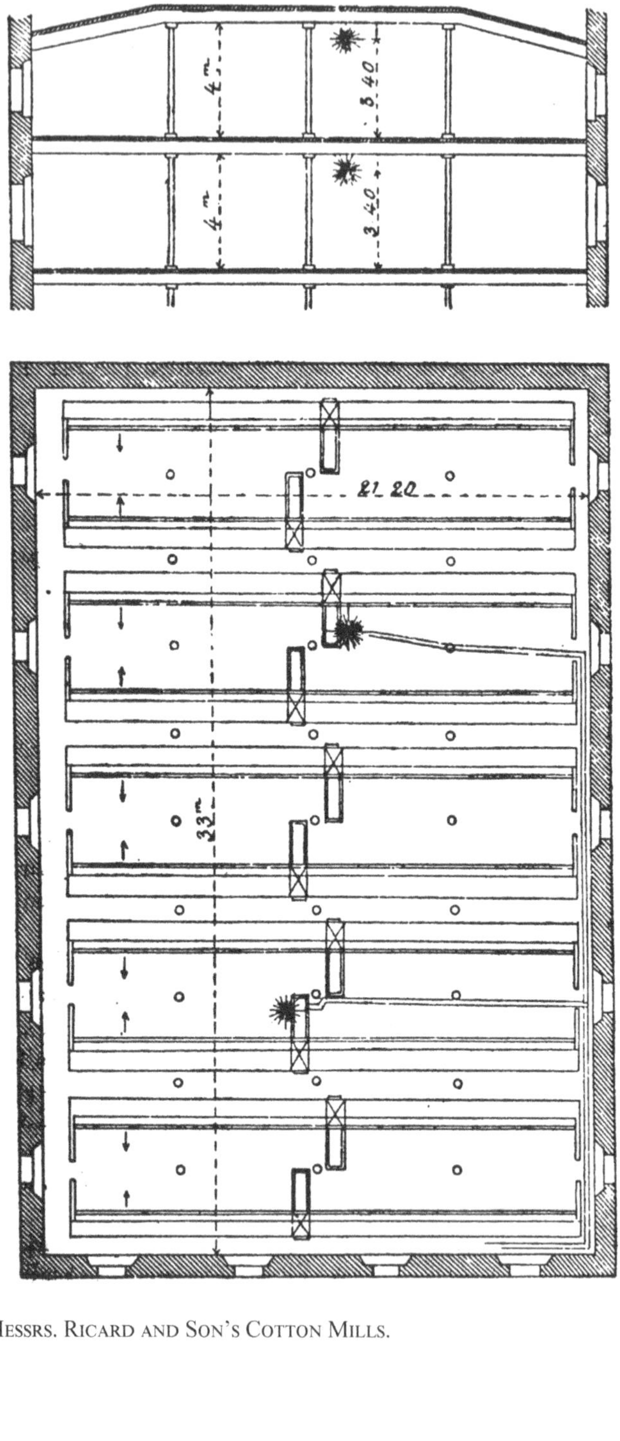

Fig. 41. MESSRS. RICARD AND SON'S COTTON MILLS.

a complete works, including building, motor, transmission, and the Gramme apparatus, it became more economical to light during the whole night, so that the interest upon expenditure might be spread over a great number of hours.

The spaces lighted included:

1. A hall, 70 mètres long by 25 mètres wide and 8 mètres high;
2. A shed, 70 mètres long by 15 mètres wide and 8 mètres high;
3. A yard, 20 mètres wide, separating the hall from the shed.

The hall is lighted by two lamps placed on the diagonals, and consequently unsymmetrically, which is very advantageous for this purpose. The lamps are elevated to 4·5 mètres from the ground and are enclosed in large square lanterns. The glasses of these lanterns are painted with zinc-white at the lower part, and to such a level that from no point in the hall can the eye perceive the voltaic arc and be thus affected. The upper part, on the contrary, is left clear, and the top of the lantern is not even glazed, whence it results that the upper luminous rays strike without any obstacle upon the ceiling and sides of the hall, thickly lime-washed, and are reflected as a soft and very uniform light.

The lighting of this hall was the principal object in view. It is very powerful, and this is necessary because of the keeping of small bales amongst the large, and of the number of tickets to be read. Writing is also necessary, to record the movement of the goods, which, arriving, are distributed and finally loaded into waggons brought up to the side of the depôt.

Everywhere in the depôt there is good light; in the office, the most distant parts of the space lighted, the small passages arranged between the voluminous bales, and despite their number; even in the bottom of the covered waggons which are loading there is sufficient light.

From a document which has been obligingly communicated with the authorisation of M. Sartiaux, engineer to the Company, we extract the following:

> "The lighting has been in operation since the 17th of January, 1877, with a daily duration of fifteen and a half hours as a mean for the first months, and variable with the season.
>
> "The great light diffused in the hall admits of the work being effected with greater celerity and less men. The economy in staff is valuated at twenty-five per cent.
>
> "Every official has no longer need to carry a hand-lantern in order to search for the bales, decipher the addresses and marks, and read the lading-bills. The work is carried on at night nearly under the same conditions as by day.
>
> "We have taken means to employ only two sheds instead of three for the service of the Company, which admits of economising the construction of a new shed.

> "Besides those direct results the electric light has given the following indirect advantages, which reduce the indemnities paid by the Company. It reduces the errors in directions and the consequent delays, the damage in loading . . . ."

Let us add that it imposes obstacles to various frauds and reduces embezzlements.

The shed is lighted by a single lamp, and this is sufficient because there are only large bales to be handled. The lantern, is similar to that in the hall. The lime-washing has been recognised as necessary here as in the other building.

The yard is lighted by the lamp from the shed, which is open on the longitudinal sides. The lighting throughout is at least as good as in the streets of Paris.

The average distance of the machines from the lamp is 80 mètres.

The expenditure is now estimated at 0·75 franc per hour and per lamp. This figure will be greatly reduced when double the number of lights are introduced, because the mechanic and the lamp-trimmer will suffice for a much more important lighting.

The cost of introduction is 25,000 francs, which includes the purchase of a steam-engine of greater force than is required. This expense would be but slightly increased by the addition of three new Gramme machines.

Detail of expenses for introduction:

A locomotive engine of 12 horse-power	7,900	francs.
Transmission, belts, &c.	4,348	,,
Building.	3,560	,,
Lantern-towers, wires, and pulleys	793	,,
Three Gramme machines	4,500	,,
Four Serrin lamps	1,800	,,
Total	23,000	francs.

Without exaggeration such a plant is worth 16,000 francs, because the transmission, the building, and the engine are 30 per cent. too dear, having regard, not to what has been done, but to what was really necessary.

Outer Port of Havre. – M. Jeanne Deslandes' Yards.

One of the most remarkable applications of lighting by electricity is that made on the building-yards of M. Jeanne Deslandes, at Havre.

This contractor is completing great works intended for the enlargement of the outer port of Havre and the creation of additional space for ships remaining in port.

These additions necessitate the demolition of considerable works, such as the curved quay and south pier as far as the sea, the sluices of the Florida dock, the walls of the counterscarp, the clearing of the ground comprised in this vast extent, and the annexation of the greater part of the Florida dock to the outer port. They will give to the outer port a width of 185 mètres instead of 90 mètres, which is its present extent. The new area, including the Florida dock, will be 21 hectares instead of 11 hectares.

For the execution of the foundations, the driving of the piles, the demolition of the old walls, &c., it is only possible to work at low water; and in order that he might use the night ebbs, M. Chéron, director of the undertaking, has installed two Gramme machines, which have worked satisfactorily for six months.

Visiting in detail the works in course of execution, on a dark, starless night, we have found that men placed at distances varying from 20 to 120 mètres from the lamps could carry on all their ordinary work without the least inconvenience. The miners were piercing the old wall at 115 mètres from the nearest light and effecting the same amount of work as during the day. A locomotive, towing ten waggons, was running on a line of 1500 mètres, bringing materials right up to the works and transporting the rubbish to the assigned sites. A steam pile-driver was in action. Masons, carpenters, navvies, &c., were executing here and there works of every kind. More than 150 workmen, on a space covering 30,000 square mètres, were working without other light than that produced by the two Gramme machines.

The lights, enclosed in lanterns upon a platform of earth 5 mètres high, were in reality 15 mètres above most of the places where the work of construction or demolition was in progress. At 115 mètres we could distinctly read a newspaper, better than if it had been lighted by a gas-burner at a distance of 5 mètres. Each lamp gave a light exceeding 500 Carcel burners.

We earnestly advise engineers and contractors for public works to visit the arrangement of M. Jeanne Deslandes; there they will find the best existing solution of the problem of lighting works at night.

Messrs. Mignon, Rouart, and Delinières' Factory at Montluçon

In their iron-tube works at Montluçon, Messrs. Mignon, Rouart, and Delinières have adopted two Gramme machines and two Serrin lamps for lighting one of the workshops. The building thus lighted is a hall 60 mètres long by 35 mètres wide. The lamps are placed at a height of 6 mètres and 31·5 mètres apart. The workmen can easily do their work and move and handle heavy pieces as well as during the day. The shadows of the columns or the machinery have so little effect that one can read even in the places where they are darkest.

In a second hall, adjacent to the first and separated from it by a wall with large openings through it, there is sufficient light from the electric lamps to enable one to read everywhere, even at a distance of 85 mètres from the lamps.

At Montluçon only Gaudoin carbons are used, these being found much better than the retort carbons.

Messrs. Mignon and Rouart, satisfied with this first application, are arranging to light completely with Gramme machines the new workshops which they have just erected on the Boulevard Voltaire at Paris.

It is complete evidence in favour of the system, that all those who have ever tried it extend its application to all places where it can be applied.

Luminous Ceiling

It often happens that a large hall is badly lighted during the day, either because it forms part of a basement or because it is surrounded by other rooms intercepting the rays of the sun. In these cases, by means of a luminous ceiling, the hall can be illuminated by a white light, soft, and in all respects equivalent to sunlight. We have done this in one of the halls of the great warehouses of the Louvre at Paris, as shown in Fig. 42.

The compartment G is above and nearly over the middle of the room to be lighted. A regulator A, balanced by the counterweight C, is put in communication with a Gramme machine driven by a portable engine and placed at 50 mètres distance from the ceiling. A plate-glass E, unsilvered and frosted, prevents the luminous rays from striking too directly into the hall and producing objectionable shadows. Four surfaces, F, forming a truncated pyramid, are lined with tin-plate, and reflect downwards a great part of the light emitted by the regulator. Two rods H support the frosted glass E. A second lamp B, resting upon a bracket J, and balanced by a counterweight D, is prepared beforehand, and replaces the lamp A when its carbons are consumed. Thus there is no interruption to the light.

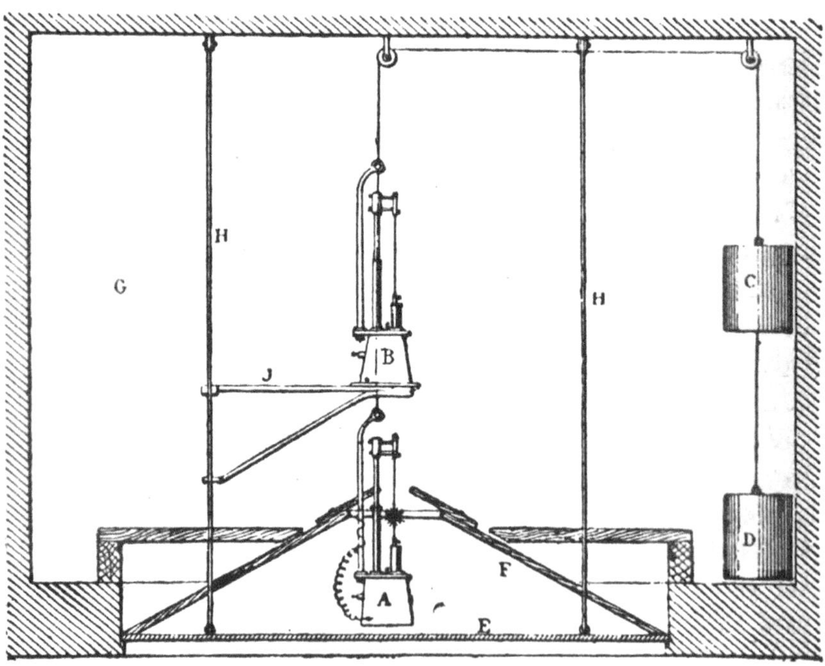

Fig. 42. Luminous Ceiling.

This arrangement was made in February, 1877, and it has not ceased to work since that time. In consequence of the temporary engine not having a very regular speed, and the carbons of the regulators not being homogeneous, there are to be observed small scintillations and sensible variations in the luminous intensity; but this inconvenience is only temporary, it will disappear completely when the proper engine is used and the Gaudoin carbons are substituted for those now employed.

The Marne and Rhine Canal at Sermaize

Electric lighting was adopted, about the end of 1875, at one of the wharves of the Marne and Rhine Canal at Sermaize, contiguous to a sugar works. It was desired to continue during the long winter nights the unloading, if then urgent, of the beetroot boats, and to take the contents immediately to the factory, without storing them on the bank of the canal, which it was important to keep quite free for traffic. This special problem was solved by a Gramme machine of 200 burners, placed beside one of the engines of the factory, at a distance of 70 mètres from the lamp. The workmen, who frequently unload two boats at the same time, can work without the least inconvenience. This has been certified by the engineer of the canal and the manager of the sugar works; the latter has written to say that he was quite satisfied with this lighting, that the Gramme machine had attained admirably the object proposed, and saved him at least 10 francs per hour.

It may be mentioned that in the sugar works the expense of electric lighting is almost limited to that of the carbon rods for the lamps, because the steam which has driven the engine can be utilised after its discharge.

Lighting a Skating-Rink

The Skating Company of Vienna (Austria) has established lighting by electricity in their large rink near Park Ring. This is the most successful open-air arrangement that has been made abroad. The surface of the ice is 5700 square mètres; the length of the space lighted, including the promenade, is 133 mètres; the total width is 57 mètres.

The light is furnished by two Gramme machines, each costing 3000 francs, placed at 135 mètres from the rink, and worked by a portable engine of 8 horse-power. The lamps are of the Serrin type. They are placed at the top of two fir poles, 7·50 mètres high. They are 57 mètres apart. Two reflectors prevent the waste of light upwards, and concentrate all the rays upon the surface of the ice. For this purpose, the eight segments which compose each of these reflectors are curved to an elliptical form, having one of the foci at the luminous point, and the other 1 mètre lower.

All parts of the space are very well lighted, and altogether the arrangements are most satisfactory.

It is to be remarked that no other system could give so good a result, because the principal condition for a skating-rink is that its centre should not be encumbered by a series of posts. It is, besides, well known that gas undergoes great variations in the open air, and that a very great expense would be involved in lighting by it a space of 5700 square mètres.

The best proof we can give of the success of this application is, that the Skating Company, who had hired the apparatus, have now purchased them.

Various Applications

We might describe many other applications, but the preceding will doubtless suffice to make the services rendered by electric lighting fully appreciated. A summary of some important applications will complete this chapter.

The cannon foundry of Bourges has eleven complete apparatus for its different workshops; Cail's manufactory at Paris has three electric lamps lighting an erecting shop of 5000 square mètres; the Mediterranean Ironworks and Dockyards Company has seven Gramme machines in its Havre workshops; the engineering works at Pantin have four complete apparatus. The works of Messrs. Crespin and Marteau, at Paris; Beaudet, at Argenteuil; Thomas and Powel, at Rouen; Ackermann, at Stockholm; Avondo, at Milan, &c., are likewise provided with Gramme machines.

We may also mention the works of Fives-Lille and of Tarbes; the engineering works of Barcelona; the Stations du Midi at Brussels; the ironworks of Fourchambault; the foundries of Bessèges and of Fumel, &c.

The quality which the electric light possesses of maintaining the tints of colours has been utilised in many manufactures, notably in the dye-works of Messrs. Guaydet and Son at Roubaix, and in those of Messrs. Hannart Brothers at Wasquehall. Messrs. Hannart have stated that they can dye pearl-greys by the electric light, and that their production has notably increased since the adoption of the Gramme machine.

We have established lighting by electricity in the workshops of M. de Quillacq, at Anzin; the weaving shop of M. Baudot, at Bar-le-Duc; in the laundries of the Lyons hospitals, &c. More than 100 new applications are ordered: a great number are now being completed.

Editorial Headnote

22. "On the Augmentation of the Power of a Magnet by the Reaction thereon of Currents Induced by the Magnet Itself," *Proceedings of the Royal Society* 15, February 14, 1867, pp. 367–372.

In the years prior to the invention and circulation of dynamos (electric generators) that could be used to sustain an electric current, scientists and tinkerers theorized a number of mechanisms for creating a constant current. Following Michael Faraday's description of electromagnetism, many "electricians" depended upon a permanent magnet, a magnet that sustains its own magnetic properties in the absence of an inducing field or current. However, a series of experiments beginning in 1866 demonstrated that permanent magnets were unnecessary. In effect, it redefined how electricians conceptualized electromagnetic apparatuses. As was typical in the British electrical community, the credit for the discovery is difficult to assign because three people, the English telegraph engineer Charles Wheatstone (1807–1875), the German engineer Werner Siemens (1816–1892), published by his brother C. W. Siemens (1823–1883), and English electrician Samuel Alfred Valery (1832–1921), independently described the new principle. Wheatstone and Siemen's papers were read together at the Royal Society and appeared in the same volume of *Philosophical Transactions*. These papers, and the discovery they describe, highlight the transnational and cooperative nature of electrical science in the 1860s. Wheatstone's paper, which he gave while demonstrating the theory, describes the phenomenon using the dynamo developed by Henry Wilde (1833–1919) and proceed to show how "weak" magnets can be used in a device to generate powerful currents. C. W. Siemens' paper describes his brother's work in Berlin. In addition to focusing on the possibility of generating a powerful current using weak magnets, Siemens celebrates a "hitherto unrecognized fact." He also notes the potential utility of this observation by showing that mechanical force can produce powerful electric currents without relying on magnets which are more liable to fail over time. Experiments like this demarcate the transition of electric power from the experimental table to the power station. Both papers cite decades-old theories and experiments by other electricians and showcase the process, rather than the moments, of discovery in electrical physics during this tumultuous period for the field.

22

ON THE AUGMENTATION OF THE POWER OF A MAGNET BY THE REACTION THEREON OF CURRENTS INDUCED BY THE MAGNET ITSELF

Charles Wheatstone

Source: Charles Wheatstone, "On the Augmentation of the Power of a Magnet by the Reaction thereon of Currents Induced by the Magnet Itself," *Proceedings of the Royal Society* 15, February 14, 1867, pp. 367–372.

The magneto-electric machines which have been hitherto described are actuated either by a permanent magnet or by an electro-magnet deriving its power from a rheomotor placed in the circuit of its coil. In the present note I intend to show that an electro-magnet, if it possess at the commencement the slightest polarity, may become a powerful magnet by the gradually augmenting currents which itself originates.

The following is a description of the form and dimensions of the electro-magnet I have employed. The construction, it will be seen, is the same as that of the electro-magnetic part of Mr. Wilde's machine.

The core of the electro-magnet is formed of a plate of soft iron 15 inches in length and ½ an inch in breadth, bent at the middle of its length into a horseshoe form. Round it is coiled in the direction of its breadth, 640 feet of insulated copper wire $\frac{1}{12}$ of an inch in diameter. The armature, which is according to Siemens's ingenious construction, consists of a rotating cylinder of soft iron 8½ inches in length, grooved at two opposite sides so as to allow the wire to be coiled upon it longitudinally; the length of the wire thus coiled is 80 feet, and its diameter is the same as that of the electro-magnet coil.

When this electro-magnet is excited by any rheomotor the current from which is in a constant direction, during the rotation of the armature currents are generated in its coil during each semirevolution, which are alternately in opposite directions; these alternate currents may be transmitted unchanged to another part of the circuit, or by means of a rheotrope be converted to the same direction.

If now, while the circuit of the armature remains completed, the rheomotor be removed from the electro-magnet, on causing the armature to revolve, however

DOI: 10.4324/9781003295556-25

rapidly, it will be found by the interposition of a galvanometer, or any other test, that but very slight effects take place. Though these effects become stronger in proportion to the residual magnetism left in the electro-magnet from the previous action of a current, they never attain any considerable amount.

But if the wires of the two circuits be so joined as to form a single circuit, in which the currents generated by the armature, after being changed to the same direction, act so as to increase the existing polarity of the electro-magnet, very different results will be obtained. The force required to move the machine will be far greater, showing a great increase of magnetic power in the horseshoe; and the existence of an energetic current in the wire is shown by its action on a galvanometer, by its heating 4 inches of platinum wire ·0067 in diameter, by its making a powerful electro-magnet, by its decomposing water, and by other tests.

The explanation of these effects is as follows: – The electro-magnet always retains a slight residual magnetism, and is therefore in the condition of a weak permanent magnet; the motion of the armature occasions feeble currents in alternate directions in the coils thereof, which, after being reduced to the same direction, pass into the coil of the electro-magnet in such manner as to increase the magnetism of the iron core; the magnet having thus received an accession of strength, produces in its turn more energetic currents in the coil of the armature; and these alternate actions continue until a maximum is attained, depending on the rapidity of the motion and the capacity of the electro-magnet.

If the two coils be connected in such manner that the rectified current from the coil of the armature passes into the coil of the electro-magnet in the direction which would impart a contrary magnetism to the iron core, no current is produced, and consequently there is no augmentation of magnetism.

It is easy to prove that the residual magnetism of the electro-magnet is the determining cause of these powerful effects. For this purpose it is sufficient to pass a current from a voltaic battery, a magneto-electric machine, or any other rheomotor, into the coil of the electro-magnet in either direction, and it will invariably be found that the direction of the current, however powerful it may eventually become, is in accordance with the polarity of the magnetism impressed on the iron core.

If, instead of the currents in the coil of the rotating armature being reduced to the same uniform direction, they retain their alternations, no effects, or at most very small differential ones, are produced, as no accumulation of magnetism then takes place.

I will now call attention to the fact that stronger effects are produced at the first moment of completing the combined circuit than afterwards. The machine having been put in motion, at the first moment of completing the circuit 4 inches of platina wire were made red-hot, but immediately afterwards the glow disappeared, and only about one inch of the wire could be permanently kept at a red heat. This diminution of effect was accompanied by a great increase of the resistance of the machine. The cause of the momentary strong effect was, that the machine from its acquired momentum continued its motion for a few seconds, though it required a stronger force than could be applied to maintain that motion. Each time the circuit is broken and recompleted the same effect recurs.

On bringing the primary coil of an inductorium (Ruhmkorff's coil) into the circuit formed by connecting the coils of the electro-magnet and rotating armature, no spark occurs in the secondary coil. On account of the great resistance of the circuit, which now also includes the primary coil of the inductorium, the current is not in sufficient quantity to produce any noticeable inductive effect.

A very remarkable increase of all the effects, accompanied by a diminution in the resistance of the machine, is observed when a cross wire is placed so as to divert a great portion of the current from the electro-magnet. The four inches of platinum wire, instead of flashing into redness and then disappearing, remains permanently ignited. The inductorium, which before gave no spark, now gave one a quarter of an inch in length; water was more abundantly decomposed; and all the other effects were similarly increased.

I account for this augmentation of the effects in the following way: – Though so much of the current is diverted from the electro-magnet by the cross wire, the magnetic effect still continues to accumulate, though not to so high a degree; but the current generated by the armature, passing through the short circuit formed by the armature-branch and cross wire, experiences a far less resistance than if it had passed through the armature and electro-magnet branches; and though the electromotive force is less, the resistance having been rendered less in a much greater proportion, the resultant effect is greater.

I must observe that a certain amount of resistance in the cross wire is necessary to produce the maximum effect. If the resistance be too small, the electro-magnet does not acquire sufficient magnetism; and if it be too great, though the magnetism becomes stronger, the increase of resistance more than counterbalances its effect.

But the effects already described are far inferior to those obtained by causing them to take place in the cross wire itself. With the same application of force, 7 inches of platinum wire were made red-hot, and sparks were elicited in the inductorium 2½ inches in length.

The force of two men was employed in these, as well as in the other experiments. When the interrupter of the primary coil was fixed, the machine was much easier to move than when it acted. For when the interrupter acted, at each moment of interruption the cross wire being, as it were, removed, the whole of the current passed through the electro-magnet, and consequently a greater amount of magnetic energy was excited, while in the intervals during which the cross wire was complete the current passed mainly through the primary coil.

The effects are much less influenced by a resistance in the electro-magnet branch than in either of the other branches.

To reduce the length of the spark in the inductorium (the primary coil of which was placed in the cross wire) to ¾ of an inch, it required the resistance of 5¼ inches of the fine platinum wire in the cross wire, 5 inches in the armature-branch, and 4 feet in the electro-magnet branch.

When there was no extra resistance in either of the branches, the length of the cross wire being only about a few feet, the intensity of the current in the electro-magnet branch, compared with that in the cross wire, was as 1: 60; and when the resistance of the primary coil of the inductorium was interposed in the cross wire, the relative intensities were as 1: 42.

In conclusion I will mention that there is an evident analogy between the augmentation of the power of a weak magnet by means of an inductive action produced by itself, and that accumulation of power shown in the static electric machines of Holtz and others which have recently excited considerable attention, in which a very small quantity of electricity directly excited is, by a series of inductive actions, augmented so as to equal, and even exceed, the effects of the most powerful machines of the ordinary construction.

Editorial Headnote

23. "A Dynamical Theory of the Electromagnetic Field," *Philosophical Transactions of the Royal Society*, 155, 1865, pp. 459–466.

James Clerk Maxwell (1831–1879) is undoubtedly one of the most impactful scientists of the nineteenth century. Born in Scotland and subsequently educated at the University of Edinburgh and Cambridge, he is often cited for his equations which unified several disparate theories in physics, and is sometimes heralded as the founder of the discipline of electrical engineering. He is perhaps best known for his experiments on the relationship between light, electricity, and magnetism, especially the publication of this paper, which describes how electricity and magnetic fields move through space as waves at the speed of light. Beginning in 1855, Maxwell, largely building on Michael Faraday's (1791–1867) work on "lines of force," began to quantify and reduce the equations used to describe the relationship between electricity and magnetism. Maxwell demonstrated that the waves his equations described could be measured and predicted. In this article he wrote that "the agreement of the results seems to show that light and magnetism are affections of the same substance." Not only did this quantify Faraday's earlier work. It helped solidify Hans Christian Ørsted's (1777–1851) argument about the existence of a relationship between electricity and magnetism from 1820 and André-Marie Ampére's (1775–1836) laws that related a magnetic field to the electric current it produced. Maxwell built a theory of electromagnetic fields. Apart from refining other theories, this branch of physics allowed electricians to describe the movement of a current, and, after Heinrich Hertz (1857–1894) proved the existence of these invisible waves, launched new branches of physics that connected directly with major twentieth-century discoveries, like the theory of relativity or the existence of subatomic particles. Although Maxwell's theory did not immediately lead to a universal absolute unit of measurement for electricity, it helped forge an alliance between telegraph engineers and mathematical physics. In addition to these specific changes to electrical physics, historians have noted that Maxwell's method of argumentation through analogy represented a shift in experimental argumentation. Even though "invisible," for Maxwell mathematical structures maintained an ontological status as any real object would. In this publication and in his earlier work on Faraday, Maxwell forged not only a new theory but a new way of thinking through analogy on paper to thinking with objects in a laboratory that dramatically shaped the direction of theoretical physics.

23

A DYNAMICAL THEORY OF THE ELECTROMAGNETIC FIELD

Maxwell James Clerk

Source: Maxwell James Clerk, "A Dynamical Theory of the Electromagnetic Field," *Philosophical Transactions of the Royal Society*, 155, 1865, pp. 459–466.

PART I. – INTRODUCTORY

(1) THE most obvious mechanical phenomenon in electrical and magnetical experiments is the mutual action by which bodies in certain states set each other in motion while still at a sensible distance from each other. The first step, therefore, in reducing these phenomena into scientific form, is to ascertain the magnitude and direction of the force acting between the bodies, and when it is found that this force depends in a certain way upon the relative position of the bodies and on their electric or magnetic condition, it seems at first sight natural to explain the facts by assuming the existence of something either at rest or in motion in each body, constituting its electric or magnetic state, and capable of acting at a distance according to mathematical laws.

In this way mathematical theories of statical electricity, of magnetism, of the mechanical action between conductors carrying currents, and of the induction of currents have been formed. In these theories the force acting between the two bodies is treated with reference only to the condition of the bodies and their relative position, and without any express consideration of the surrounding medium.

These theories assume, more or less explicitly, the existence of substances the particles of which have the property of acting on one another at a distance by attraction or repulsion. The most complete development of a theory of this kind is that of M. W. WEBER,[1] who has made the same theory include electrostatic and electromagnetic phenomena.

In doing so, however, he has found it necessary to assume that the force between two electric particles depends on their relative velocity, as well as on their distance.

This theory, as developed by MM. W. WEBER and C. NEUMANN,[2] is exceedingly ingenious, and wonderfully comprehensive in its application to the phenomena of statical electricity, electromagnetic attractions, induction of currents

DOI: 10.4324/9781003295556-26

and diamagnetic phenomena; and it comes to us with the more authority, as it has served to guide the speculations of one who has made so great an advance in the practical part of electric science, both by introducing a consistent system of units in electrical measurement, and by actually determining electrical quantities with an accuracy hitherto unknown.

(2) The mechanical difficulties, however, which are involved in the assumption of particles acting at a distance with forces which depend on their velocities are such as to prevent me from considering this theory as an ultimate one, though it may have been, and may yet be useful in leading to the coordination of phenomena.

I have therefore preferred to seek an explanation of the fact in another direction, by supposing them to be produced by actions which go on in the surrounding medium as well as in the excited bodies, and endeavouring to explain the action between distant bodies without assuming the existence of forces capable of acting directly at sensible distances.

(3) The theory I propose may therefore be called a theory of the *Electromagnetic Field,* because it has to do with the space in the neighbourhood of the electric or magnetic bodies, and it may be called a *Dynamical* Theory, because it assumes that in that space there is matter in motion, by which the observed electromagnetic phenomena are produced.

(4) The electromagnetic field is that part of space which contains and surrounds bodies in electric or magnetic conditions.

It may be filled with any kind of matter, or we may endeavour to render it empty of all gross matter, as in the case of GEISSLER's tubes and other so-called vacua.

There is always, however, enough of matter left to receive and transmit the undulations of light and heat, and it is because the transmission of these radiations is not greatly altered when transparent bodies of measurable density are substituted for the so-called vacuum, that we are obliged to admit that the undulations are those of an æthereal substance, and not of the gross matter, the presence of which merely modifies in some way the motion of the æther.

We have therefore some reason to believe, from the phenomena of light and heat, that there is an æthereal medium filling space and permeating bodies, capable of being set in motion and of transmitting that motion from one part to another, and of communicating that motion to gross matter so as to heat it and affect it in various ways.

(5) Now the energy communicated to the body in heating it must have formerly existed in the moving medium, for the undulations had left the source of heat some time before they reached the body, and during that time the energy must have been half in the form of motion of the medium and half in the form of elastic resilience. From these considerations Professor W. THOMSON has argued,[3] that the medium must have a density capable of comparison with that of gross matter, and has even assigned an inferior limit to that density.

(6) We may therefore receive, as a datum derived from a branch of science independent of that with which we have to deal, the existence of a pervading

medium, of small but real density, capable of being set in motion, and of transmitting motion from one part to another with great, but not infinite, velocity.

Hence the parts of this medium must be so connected that the motion of one part depends in some way on the motion of the rest; and at the same time these connexions must be capable of a certain kind of elastic yielding, since the communication of motion is not instantaneous, but occupies time.

The medium is therefore capable of receiving and storing up two kinds of energy, namely, the "actual" energy depending on the motions of its parts, and "potential" energy, consisting of the work which the medium will do in recovering from displacement in virtue of its elasticity.

The propagation of undulations consists in the continual transformation of one of these forms of energy into the other alternately, and at any instant the amount of energy in the whole medium is equally divided, so that half is energy of motion, and half is elastic resilience.

(7) A medium having such a constitution may be capable of other kinds of motion and displacement than those which produce the phenomena of light and heat, and some of these may be of such a kind that they may be evidenced to our senses by the phenomena they produce.

(8) Now we know that the luminiferous medium is in certain cases acted on by magnetism; for FARADAY[4] discovered that when a plane polarized ray traverses a transparent diamagnetic medium in the direction of the lines of magnetic force produced by magnets or currents in the neighbourhood, the plane of polarization is caused to rotate.

This rotation is always in the direction in which positive electricity must be carried round the diamagnetic body in order to produce the actual magnetization of the field.

M. VERDET[5] has since discovered that if a paramagnetic body, such as solution of perchloride of iron in ether, be substituted for the diamagnetic body, the rotation is in the opposite direction.

Now Professor W. THOMSON[6] has pointed out that no distribution of forces acting between the parts of a medium whose only motion is that of the luminous vibrations, is sufficient to account for the phenomena, but that we must admit the existence of a motion in the medium depending on the magnetization, in addition to the vibratory motion which constitutes light.

It is true that the rotation by magnetism of the plane of polarization has been observed only in media of considerable density; but the properties of the magnetic field are not so much altered by the substitution of one medium for another, or for a vacuum, as to allow us to suppose that the dense medium does anything more than merely modify the motion of the ether. We have therefore warrantable grounds for inquiring whether there may not be a motion of the ethereal medium going on wherever magnetic effects are observed, and we have some reason to suppose that this motion is one of rotation, having the direction of the magnetic force as its axis.

(9) We may now consider another phenomenon observed in the electromagnetic field. When a body is moved across the lines of magnetic force it experiences what is called an electromotive force; the two extremities of the body tend to become oppositely electrified, and an electric current tends to flow through the body. When the electromotive force is sufficiently powerful, and is made to act on certain compound bodies, it decomposes them, and causes one of their components to pass towards one extremity of the body, and the other in the opposite direction.

Here we have evidence of a force causing an electric current in spite of resistance; electrifying the extremities of a body in opposite ways, a condition which is sustained only by the action of the electromotive force, and which, as soon as that force is removed, tends, with an equal and opposite force, to produce a counter current through the body and to restore the original electrical state of the body; and finally, if strong enough, tearing to pieces chemical compounds and carrying their components in opposite directions, while their natural tendency is to combine, and to combine with a force which can generate an electromotive force in the reverse direction.

This, then, is a force acting on a body caused by its motion through the electromagnetic field, or by changes occurring in that field itself; and the effect of the force is either to produce a current and heat the body, or to decompose the body, or, when it can do neither, to put the body in a state of electric polarization, – a state of constraint in which opposite extremities are oppositely electrified, and from which the body tends to relieve itself as soon as the disturbing force is removed.

(10) According to the theory which I propose to explain, this "electromotive force" is the force called into play during the communication of motion from one part of the medium to another, and it is by means of this force that the motion of one part causes motion in another part. When electromotive force acts on a conducting circuit, it produces a current, which, as it meets with resistance, occasions a continual transformation of electrical energy into heat, which is incapable of being restored again to the form of electrical energy by any reversal of the process.

(11) But when electromotive force acts on a dielectric it produces a state of polarization of its parts similar in distribution to the polarity of the parts of a mass of iron under the influence of a magnet, and like the magnetic polarization, capable of being described as a state in which every particle has its opposite poles in opposite conditions.[7]

In a dielectric under the action of electromotive force, we may conceive that the electricity in each molecule is so displaced that one side is rendered positively and the other negatively electrical, but that the electricity remains entirely connected with the molecule, and does not pass from one molecule to another. The effect of this action on the whole dielectric mass is to produce a general displacement of electricity in a certain direction. This displacement does not amount to a current, because when it has attained to a certain value it remains constant, but it is the commencement of a current, and its variations constitute currents in the positive or the negative direction according as the displacement is increasing or decreasing. In the interior of the dielectric there is no indication of electrification, because the electrification of the surface of any molecule is neutralized by the opposite electrification of the surface of the molecules in contact with it; but at the bounding surface of the dielectric, where the electrification is not neutralized, we find the phenomena which indicate positive or negative electrification.

The relation between the electromotive force and the amount of electric displacement it produces depends on the nature of the dielectric, the same electromotive force producing generally a greater electric displacement in solid dielectrics, such as glass or sulphur, than in air.

(12) Here, then, we perceive another effect of electromotive force, namely, electric displacement, which according to our theory is a kind of elastic yielding to the action of thé force, similar to that which takes place in structures and machines owing to the want of perfect rigidity of the connexions.

(13) The practical investigation of the inductive capacity of dielectrics is rendered difficult on account of two disturbing phenomena. The first is the conductivity of the dielectric, which, though in many cases exceedingly small, is not altogether insensible. The second is the phenomenon called electric absorption,[8] in virtue of which, when the dielectric is exposed to electromotive force, the electric displacement gradually increases, and when the electromotive force is removed, the dielectric does not instantly return to its primitive state, but only discharges a portion of its electrification, and when left to itself gradually acquires electrification on its surface, as the interior gradually becomes depolarized. Almost all solid dielectrics exhibit this phenomenon, which gives rise to the residual charge in the Leyden jar, and to several phenomena of electric cables described by Mr. F. JENKIN.[9]

(14) We have here two other kinds of yielding besides the yielding of the perfect dielectric, which we have compared to a perfectly elastic body. The yielding due to conductivity may be compared to that of a viscous fluid (that is to say, a fluid having great internal friction), or a soft solid on which the smallest force produces a permanent alteration of figure increasing with the time during which the force acts. The yielding due to electric absorption may be compared to that of a cellular elastic body containing a thick fluid in its cavities. Such a body, when subjected to pressure, is compressed by degrees on account of the gradual yielding of the thick fluid; and when the pressure is removed it does not at once recover its figure, because the elasticity of the substance of the body has gradually to overcome the tenacity of the fluid before it can regain complete equilibrium.

Several solid bodies in which no such structure as we have supposed can be found, seem to possess a mechanical property of this kind[10]; and it seems probable that the same substances, if dielectrics, may possess the analogous electrical property, and if magnetic, may have corresponding properties relating to the acquisition, retention, and loss of magnetic polarity.

(15) It appears therefore that certain phenomena in. electricity and magnetism lead to the same conclusion as those of optics, namely, that there is an æthereal

medium pervading all bodies, and modified only in degree by their presence; that the parts of this medium are capable of being set in motion by electric currents and magnets; that this motion is communicated from one part of the medium to another by forces arising from the connexions of those parts; that under the action of these forces there is a certain yielding depending on the elasticity of these connexions; and that therefore energy in two different forms may exist in the medium, the one form being the actual energy of motion of its parts, and the other being the potential energy stored up in the connexions, in virtue of their elasticity.

(16) Thus, then, we are led to the conception of a complicated mechanism capable of a vast variety of motion, but at the same time so connected that the motion of one part depends, according to definite relations, on the motion of other parts, these motions being communicated by forces arising from the relative displacement of the connected parts, in virtue of their elasticity. Such a mechanism must be subject to the general laws of Dynamics, and we ought to be able to work out all the consequences of its motion, provided we know the form of the relation between the motions of the parts.

(17) We know that when an electric current is established in a conducting circuit, the neighbouring part of the field is characterized by certain magnetic properties, and that if two circuits are in the field, the magnetic properties of the field due to the two currents are combined. Thus each part of the field is in connexion with both currents, and the two currents are put in connexion with each other in virtue of their connexion with the magnetization of the field. The first result of this connexion that I propose to examine, is the induction of one current by another, and by the motion of conductors in the field.

The second result, which is deduced from this, is the mechanical action between conductors carrying currents. The phenomenon of the induction of currents has been deduced from their mechanical action by HELMHOLTZ[11] and THOMSON.[12] I have followed the reverse order, and deduced the mechanical action from the laws of induction. I have then described experimental methods of determining the quantities L, M, N, on which these phenomena depend.

(18) I then apply the phenomena of induction and attraction of currents to the exploration of the electromagnetic field, and the laying down systems of lines of magnetic force which indicate its magnetic properties. By exploring the same field with a magnet, I show the distribution of its equipotential magnetic surfaces, cutting the lines of force at right angles.

In order to bring these results within the power of symbolical, calculation, I then express them in the form of the General Equations of the Electromagnetic Field. These equations express –

(A) The relation between electric displacement, true conduction, and the total current, compounded of both.
(B) The relation between the lines of magnetic force and the inductive coefficients of a circuit, as already deduced from the laws of induction.

(C) The relation between the strength of a current and its magnetic effects, according to the electromagnetic system of measurement.
(D) The value of the electromotive force in a body, as arising from the motion of the body in the field, the alteration of the field itself, and the variation of electric potential from one part of the field to another.
(E) The relation between electric displacement, and the electromotive force which produces it.
(F) The relation between an electric current, and the electromotive force which produces it.
(G) The relation between the amount of free electricity at any point, and the electric displacements in the neighbourhood.
(H) The relation between the increase or diminution of free electricity and the electric currents in the neighbourhood.

There are twenty of these equations in all, involving twenty variable quantities.

(19) I then express in terms of these quantities the intrinsic energy of the Electromagnetic Field as depending partly on its magnetic and partly on its electric polarization at every point.

From this I determine the mechanical force acting, 1st, on a moveable conductor carrying an, electric current; 2ndly, on a magnetic pole; 3rdly, on an electrified body.

The last result, namely, the mechanical force acting on an electrified body, gives rise to an independent method of electrical measurement founded on its electrostatic effects. The relation between the units employed in the two methods is shown to depend on what I have called the "electric elasticity" of the medium, and to be a velocity, which has been experimentally determined by MM. WEBER and KOHLRAUSCH.

I then show how to calculate the electrostatic capacity of a condenser, and the specific inductive capacity of a dielectric.

The case of a condenser composed of parallel layers of substances of different electric resistances and inductive capacities is next examined, and it is shown that the phenomenon called electric absorption will generally occur, that is, the condenser, when suddenly discharged, will after a short time show signs of a *residual* charge.

(20) The general equations are next applied to the case of a magnetic disturbance propagated through a non-conducting field, and it is shown that the only disturbances which can be so propagated are those which are transverse to the direction of propagation, and that the velocity of propagation is the velocity v, found from experiments such as those of WEBER, which expresses the number of electrostatic units of electricity which are contained in one electromagnetic unit.

This velocity is so nearly that of light, that it seems we have strong reason to conclude that light itself (including radiant heat, and other radiations if any) is an electromagnetic disturbance in the form of waves propagated through the electromagnetic field according to electromagnetic laws. If so, the agreement between the elasticity of the medium as calculated from the rapid alternations of luminous vibrations, and as found by the slow processes of electrical experiments, shows how perfect and regular the elastic properties of the medium must be when not encumbered with any matter denser than air. If the same character of the elasticity is retained in dense transparent bodies, it appears that the square of the index of refraction is equal to the product of the specific dielectric capacity and the specific magnetic capacity. Conducting media are shown to absorb such radiations rapidly, and therefore to be generally opaque.

The conception of the propagation of transverse magnetic disturbances to the exclusion of normal ones is distinctly set forth by Professor FARADAY[13] in his "Thoughts on Ray Vibrations." The electromagnetic theory of light, as proposed by him, is the same in substance as that which I have begun to develope in this paper, except that in 1846 there were no data to calculate the velocity of propagation.

(21) The general equations are then applied to the calculation of the coefficients of mutual induction of two circular currents and the coefficient of self-induction in a coil. The want of uniformity of the current in the different parts of the section of a wire at the commencement of the current is investigated, I believe for the first time, and the consequent correction of the coefficient of self-induction is found.

These results are applied to the calculation of the self-induction of the coil used in the experiments of the Committee of the British Association on Standards of Electric Resistance, and the value compared with that deduced from the experiments.

Source Endnotes

1 Electrodynamische Maassbestiinmimgen. Leipzic Trans, vol. i. 1849, and TAYLOR's Scientific Memoirs, vol. v. art. xiv.

2 "Explicate tentatur quomodo fiat ut lucis planum polarizationis per vires electrieas vel magneticas declinetur." – Halis Saxonum, 1858.

3 "On the Possible Density of the Luminiferotis Medium, and on the Mechanical Value of a Cubic Mile of Sunlight," Transactions of the Royal Society of Edinburgh (1854), p. 57.

4 Experimental Researches, Series 19.

5 Comptes Rendus (1856, second half year, p. 529, and 1857, first half year, p. 1209).

6 Proceedings öf the Royal Society, June 1856 and June 1861.

7 FARADAY Exp. Res. Series XI.; MOSSOTTI, Mem. della Soc. Italiana (Modena), vol. xxiv. part 2. p. 49.

8 FARADAY, Exp. Bes. 1233–1250.

9 Reports of British. Association 1859, p. 248; and Report of Committee of Board of Trade on Submarine Cables, pp. 136 & 464.

10 As, for instance, the composition of glue, treacle, &c., of which small plastic figures are made, which after being distorted gradually recover their shape.

11 "Conservation of Force," Physical Society of Berlin, 1847; and TAYLOR's Scientific Memoirs, 1853, p. 114.

12 Reports of the British Association, 1848; Philosophical Magazine, Dec. 1851.

13 Philosophical Magazine, May 1846, or Experimental Researches, iii. p. 447.

Editorial Headnote

24. "Aron Meter," *The Telegraphic Journal and Electrical Review*, February 7, 1890, pp. 158–159.

In 1883, Hermann Aron patented the "pendelzähler," or watt-hour meter, that contained two clocks wrapped in coils that measured watts over time. Over the course of that decade the meter, often known as the "Aronometer," gained wider circulation as electric power became a commodity to measure, and of course sell. In the *Telegraphic Journal and Electrical Review*, a major electrical science and engineering publication, by 1890, Gisbert Johann Eduard Kapp (1852–1922) explored the utility and accuracy of the device in a letter to General Electric Company, Ltd. Kapp was an Austrian-English electrical engineer who gained notice as chair of electrical engineering department at the University of Birmingham and served as President of the Institution of Electrical Engineers. Beyond his credentials, Kapp is a marvelous example of the British and Continental connections around electrical engineering through invention, publication, and education. In the letter, Kapp explains 10 days' worth of tests he ran on the Aronometer the company had sent him, in which he tested the device with lamps and alternating and direct currents to determine the accuracy of the device. After comparing the device with other meters, such as William Thomson's electrostatic voltmeter, Kapp concluded that with minor errors the meter would function well for practical purposes, by which he certainly meant the systemic measure of power consumption and use. The letter highlights the early extension of electrical experimentation into the service of utilities like General Electric.

24

ARON METER

Gisbert Kapp

Source: Gisbert Kapp, "Aron Meter," *The Telegraphic Journal and Electrical Review*, February 7, 1890, pp. 158–159.

SCIENTIFIC reports on new inventions are sometimes viewed with a little suspicion, especially when the invention has not had the advantage of prolonged public use. Dr. Aron's electricity meter is, however, not a new invention, and it has been employed for a number of years, there being several thousands working daily in England and on the Continent.

Considering its almost universal use, particularly on the Continent, it seems strange that no independent and complete testimony has found its way into print, therefore we deem it all the more desirable to reproduce the results of careful tests made by Mr. Gisbert Kapp, which demonstrate conclusively the accuracy of the Aron meter, both for alternating and continuous currents. Experience leads Mr. Kapp to the conclusion that the accuracy, long range and economy of power, make it a very satisfactory instrument, both for direct and alternating currents.

27th January, 1890.

Messrs. the General Electric Company, Limited,
71, Queen Victoria Street, E.C.

Gentlemen, – I have subjected the Aron meter, which you supplied, to a series of tests extending over about ten days, and now report to you on the results of these tests.

The meter consists of a pair of pendulum clocks so geared together as to register the difference in their rates on a set of dials. One of the clocks has an ordinary pendulum, whilst the bob of the pendulum of the other clock consists of a fine wire solenoid, placed horizontally and swinging within a fixed solenoid of much stouter wire, through which passes the supply current which is to be metered. The fine wire solenoid is connected in series with a resistance coil, and the two form a shunt to the supply leads. The connections are so arranged that there is attraction between the shunt solenoid and the main solenoid, and consequently acceleration of the electric as compared with the gravity pendulum; the difference in rate, which is registered on the dials, being very nearly proportional to the amount of energy passed through the meter in any given time. To find this energy we have

DOI: 10.4324/9781003295556-27

therefore to multiply the numbers registered on the dials by a coefficient, which must be determined for each meter by experiment, and which in an absolutely perfect meter should be a constant for all currents and pressures within the range of this instrument. I do not know any meter in which this coefficient is absolutely constant: it varies with the amount of current and the amount of variation is an indication of the greater or lesser accuracy of the meter. To judge of the merits of a meter it is therefore only necessary to determine the coefficient for different currents and pressures and to compare the coefficients with each other. The more nearly the coefficients agree the greater is the degree of accuracy attained by the meter.

I have tested your Aron meter at pressures varying between 180 and 95 volts, and with currents varying between 26 and 7 ampères, both direct and alternating. In addition, I have also made a prolonged test with a direct current of 0·65 ampères.

The meter is designed for a mean pressure of 100 volts and a maximum current of 26 ampères. I found that the co-efficient is practically unaffected by variations in pressure between the limits of 95 and 130 volts. The pressure was measured by a Cardew voltmeter, and by a Sir William Thomson electrostatic voltmeter. I calibrated the Cardew voltmeter by means of a Clark standard cell and its indications agreed perfectly with Sir William Thomson's instrument. The currents between 7 and 26 ampères were measured by a Sir William Thomson ampere balance, and the current of 0·65 ampères was measured by a Sir William Thomson deciampère balance. With the larger currents the readings were taken every quarter of an hour, and with the smaller currents they were taken at longer intervals. In this case the current was taken through a single lamp from a secondary battery, ordinarily supplying 40 lamps and was, therefore, absolutely constant over, a period of many hours, rendering it unnecessary to take frequent readings. The tests at high voltage (from 110 to 180 volts) were made whilst charging the secondary battery, and those at normal voltage (from 95 to 105 volts) were made whilst supplying the lamps from the battery, when the current was direct; and whilst supplying lamps from transformer, when the current was alternating. The conditions of the tests were, therefore, those which obtain in the actual use of meters.

The co-efficient given by Dr. Aron for this meter is 1·07, that is to say, at about 100 volts pressure between the mains one division in the uppermost dial represents 107 Watt hours. I first tested the meter in exactly the condition in which it came to me and found that for currents varying between 25 and 15 ampères, the co-efficient given by the maker was nearly correct. For smaller currents the co-efficient was, however, incorrect and the error increased as the current was diminished. Thus with current required for one single lamp my test gave co-efficient of only 0·97. I found that this discrepancy was due to the fact that the meter had not been properly regulated before it was sent to me. In order that a clock meter of this kind may register accurately, it is necessary that the gravity pendulum should have exactly the same period as the electric pendulum when no current is passing. In this meter the gravity pendulum had a longer period, and consequently the meter registered, even when no current was passing. This defect can very easily be rectified, since the gravity pendulum is provided with a weight which can easily be screwed up or down as may be required, in order to bring the two pendulums to absolute synchronism. I found it very easy to make this regulation, and my subsequent tests were made with the meter so regulated. I mention this matter to draw your attention to the necessity of regulating every meter after it is installed.

Test of Meter with Direct Currents. – The meter being intended for a 40-light installation, it seemed advisable to test it within the limits of current which would ordinarily obtain in such an installation. I made, therefore, prolonged tests with currents of 8 and 23 ampères, which corresponded to 12 and 35 lamps respectively. The coefficients I obtained were the following: –

With 35 lamps, 1·130; with 12 lamps, 1·141.

I also made a prolonged test with a single lamp, in order to determine the accuracy of the meter at the lowest current at which it can possibly be used in practice. This test is for all meters the most severe, since any error, however small, absolutely becomes relatively large if the current to be metered is itself very small. I found that even at the lowest limit of current the Aron meter gave a remarkably accurate result, the coefficient being:

With 1 lamp. ... ... 1·146.

On the basis of these figures it is now possible to form an estimate of the accuracy of the meter when in practical use. Taking the mean between the coefficients for 12 and 35 lamps, I find 1·1355 as the coefficient to be used with this meter. The maximum error which would obtain if the meter were always worked at the lower or always at the higher limit of current, would then be ½ per cent.; but as in practice the meter will generally be worked at intermediate currents, the probable error will be less than ½ per cent. I need hardly say that this degree of accuracy is greater than that of any gas meter, and more than sufficient for all practical purposes.

Test of Meter with Alternating Currents. – As there must necessarily be some self-induction in the fine wire coil forming the bob of the electric pendulum, the shunt current must lag behind the main current, and the acceleration will be somewhat less than with a direct current representing the same energy. It will also be evident that the difference will be the greater the higher the frequency of the alternating current, so that the coefficient of the meter, as found by experiment with direct currents, will not be applicable if the meter is to be used with alternating currents, and a new coefficient must be experimentally determined for every frequency. I have accordingly tested the meter at two frequencies which may be considered limiting values for practical work, viz., 105 cycles per second (or 12,600 reversals per minute), and 70 cycles per second (or 8,400 reversals per minute). As regards current strength, I have in both cases made the tests within the same limits as those I adopted for the direct currents, and I give the results in the following table: –

Number of lamps supplied.		Coefficient for a frequency of 105		70
35	...	1·195	...	1·170
12	...	1·220	...	1·194

Taking, as before, the averages between these figures, I find that the coefficient to be used with this méter is: –

1·207 for a frequency of 105, and
1·182 for a frequency of 70.

In either case, the maximum possible error is about 1¼ per cent, whilst the probable average error will be considerably smaller than 1¼ per cent. This degree of accuracy, though not so great as that found with direct currents, is still quite sufficient for all practical purposes.

The energy absorbed in the meter varies with current passing through it. At a pressure of 100 volts the energy absorbed is:

With 1 lamp.	... ...	3 watts.
With 12 lamps	... ...	4 "
With 35 lamps	... ...	9 "

My experiments lead me to conclude that the accuracy, long range, and economy of power shown by the Aron meter make it a very satisfactory instrument, both for direct and alternating currents. – Yours faithfully,

(signed) GISBERT KAPP.

Part 4

SELF-REFLECTION-RETROSPECTION "LOOKING"

Editorial Headnote

25. "Priority," *The Electrician* May 18, 1894, pp. 76–77.

The anonymous article "Priority," published in *The Electrician* on May 18, 1894, delves into the significance of precedence and recognition in the rapidly evolving field of electrical engineering. This piece highlights the competitive nature of technological advancements and the importance placed on establishing priority for inventions and discoveries. Connecting this discussion to the broader history of electrification in Victorian Britain, we see how the race for innovation not only propelled technological progress but also shaped the trajectory of the nation's industrial and social development.

During the late nineteenth century, Victorian Britain was at the forefront of the electrification movement, with numerous engineers and inventors striving to make groundbreaking contributions. The article "Priority" underscores the fierce competition among these pioneers, with each vying for acknowledgment and credit. This competitive spirit was a driving force behind many significant advancements in electrical technology, as inventors were motivated to push the boundaries of what was possible.

The emphasis on priority and recognition in *The Electrician* reflects broader themes in the history of electrification in Victorian Britain. The period was marked by rapid innovation and the establishment of foundational technologies that would transform society. For instance, the development of efficient generators and transformers, as well as advancements in electrical distribution systems, were critical in making electricity a viable and widespread resource. The competitive atmosphere, as highlighted in the article, spurred engineers to refine and perfect these technologies, leading to the electrification of cities and towns across Britain.

Electric lighting is a prime example of how this race for priority influenced everyday life. The transition from gas to electric lighting revolutionized urban environments, extending productive hours and improving safety. Innovators like Joseph Swan and Thomas Edison were central figures in this transformation, each claiming priority for the invention of the incandescent lightbulb. Their rivalry exemplifies the broader competitive landscape and its role in accelerating technological adoption.

Moreover, the development of electric communication systems, such as the telegraph and telephone, was heavily influenced by the quest for priority. Inventors like Samuel Morse and Alexander Graham Bell became renowned for their contributions, which facilitated instant communication across long distances. These technologies were instrumental in maintaining the cohesion of the British Empire, allowing rapid information exchange and efficient administration.

In conclusion, the article "Priority" in *The Electrician* captures a crucial aspect of the history of electrification in Victorian Britain – the competitive drive for recognition and its impact on technological progress. This relentless pursuit of innovation not only led to significant advancements in electrical engineering but also played a vital role in shaping the industrial and social landscape of the era. The story of electrification in Victorian Britain is thus one of both individual ambition and collective transformation, driven by the desire to achieve and be recognized for groundbreaking achievements.

25

PRIORITY

Anon

Source: Anon., "Priority," *The Electrician* May 18, 1894, pp. 76–77.

SOME little time ago a New York contemporary published a series of articles from the pen of Miss MARY A. HENRY – extracts from which appeared in these columns at the time – setting forth the bases of her father's claim to be called the first and true discoverer of "magneto-electricity." In attaching such paramount importance to mere priority, Miss HENRY did no more than give her adhesion to the widely-accepted and influentially-supported doctrine that questions of the kind should be settled solely by the date of publication, without, as HELMHOLTZ says, "considering the ripeness of the research."

Encouraged, probably, by the success of her former literary effort, Miss HENRY has more recently undertaken the task of assigning to JOSEPH HENRY his proper place in the history of the electro-magnetic telegraph. The place allotted is, it is, perhaps, needless to say, the first. A perusal of this second series of articles, now appearing in the *Electrical Engineer* of New York, would, no doubt, show that this second claim has been made good. But the discovery of the "intensity" electro-magnet is in a very different category to that of the phenomena connected with the interactions of electric and magnetic circuits. The "intensity" electro-magnet is the very soul of the electric telegraph, and was recognised as such from the commencement. Had the Princeton Professor been possessed of the modern mania for patenting the results of his laboratory investigations, the files of the United States Patent Office would have been enriched by at least one intrinsically valuable invention, and the patentee possibly by an untold tale of dollars. The importance of priority the moment a philosopher descends to inventing is obvious; it is the all-in-all, the alpha and omega of the trade, and cannot be over-estimated. The inventor's attachment to priority has, in short, a purely business basis, and need not, therefore, be objected to as irrational. The pocket argument, however, does not apply to the discoveries of pure science. The hot haste of the patentee we can understand, not so the pot-hunting of the physicist. Whence comes this feverish eagerness to be first, which is so characteristic of latter-day research, always to its detriment, and often to the delight of the Philistine, ever glad to see the philosopher come down from his pedestal and display to the world at large a few of the failings of common men? Does it come from that "rage for fame" which

DOI: 10.4324/9781003295556-29

"attends both great and small," and makes men think it "better to be damned than mentioned not at all"?

Now the articles which dealt with the conflicting claims of MICHAEL FARADAY and JOSEPH HENRY, albeit in excellent taste, the author refusing to allow her close relationship to one of the candidates to intervene at the expense of her fairness and courtesy, left the impression that the true significance of each of the many factors that go to make up the ultimate value of a scientific discovery had been somewhat misapprehended; a failing, indeed, common to most literature of the kind. Stress was rightly laid upon the widely different conditions under which FARADAY and HENRY worked: the one in a well-equipped laboratory, with ample leisure, and amidst books and stimulating society; the other only able to devote a few weeks every summer to research, and living out of the world in a small town 150 miles from New York. But the real objective of Miss HENRY's incursion into the fields of technical literature, the idea with which proof was piled on the top of proof *ad demonstrandum* and *ad nauseam,* was to secure a verdict for JOSEPH HENRY – to make it known unto all whom it might concern that the dark horse away out in the American backwoods had passed the winning post at least half a head in front of MICHAEL FARADAY. In the investigator himself a wish to be first may perhaps be excused by reason of the inherent defects of the human mechanism, but in settling the precise position of a niche in the Temple of Scientific Fame, the undue weight that is attached to mere matters of date is wholly without excuse. It has no foundation in reason, and should have none in sentiment. The assiduity of the scientist-historiographer tends greatly to foster this belief in the supreme importance of being first. A day amongst the dry-as-dust files of the Patent Office, a visit to the British Museum library, the possession of a moderately-sized collection of old volumes, and you have all the ingredients necessary for the unearthing of the real genuine No. 1, who somewhere in the dark ages of Science was the only son of a widowed mother, learnt three languages by sitting up late at night, and imbibed science in lieu of beer during the dinner hour. If this exhumed No. 1 be finally admitted within the pale of science, after due discussion and examination of his claims, the names of the discoverer of the discoverer and of the discoverer of the discovery will in all probability have become so linked together that, like twin stars, they will for ever revolve around each other in the scientific firmament, and be reckoned of equal magnitude.

About two things, and two things only, should history concern itself in dealing with the personal equation of a scientific discovery, and these things are its intrinsic value and the manner in which it was made. As it is, intrinsic worth counts, indeed, for something; but opportuneness, date, and mode of publication, and practical possibilities count for much more. That one man had a clearer idea of what he was aiming at, or a more intelligent perception of the true leaning and bearing of what he had found, will avail him but little in face of the fact that someone else stumbled upon the same object, it may be, quite four-and-twenty hours before. It is the knowledge of this that leads so many to sacrifice thoroughness and lucid exposition in their desire to "take date." Hence comes the modern custom of publishing, step by step, the blind gropings of the laboratory: – "Certain experiments which I have been making on the electrolysis of aniline have casually led me to suspect that there is a fundamental flaw in the Newtonian theory of gravitation; more anon (April 1st). Further experiments have convinced me of the absolute correctness of my conclusion; I have now only to test the accuracy of my electrostatic voltmeters, &c . . . . Regret to say that a redetermination of *H* and *g* in my laboratory has revealed grave inaccuracies, vitiating conclusion jumped to on April 1st, so can only take date for my 'suspicion.'"

Notwithstanding this rage for taking date, two and two, we doubt not, will continue to the end of time to make four, and the truth to be that the honour awarded to a discoverer ought in no way to be lessened, because, totally unknown to him, someone else has carried out an investigation on parallel lines, and has anticipated his conclusions. The intellectual triumph is the same in either case, and the waste of thought and experimental skill are due to the defects of the system. It does not follow that a man is entitled to "all the honours of war" who shuts himself up, and deliberately ignoring the work of his scientific ancestors and contemporaries, magnanimously gives to the world results long since common property. An indispensable preliminary to the solution of every scientific problem ought to be to make oneself acquainted with the results of previous work in the same field, but in these days of verbiage and voluminousness an investigator who would exhaust the literature of his subject before commencing his own attack runs a risk of never beginning – of not seeing the forest for the trees. After all, to quote from HELMHOLTZ once more, "All this literature of untried and unconfirmed hypotheses has no value in the progress of science. On the contrary, the few sound ideas which it may contain are concealed by the rubbish of the rest; and one who wants to publish something really new – facts – sees himself open to the danger of countless claims of priority unless he is prepared to waste time and power in reading beforehand a quantity of absolutely useless books, and to destroy his readers' patience by a multitude of useless quotations." In the past a lack of proper means of intercommunication led to much wasteful working on parallel lines. To-day the danger is that this absurd ambition to publish first "without considering the ripeness of the research," in order to secure the equivocal advantage of priority, leading as it does to the publication of an immense amount of "untried and unconfirmed hypotheses," will equally result in wasteful working on parallel lines. The volume of the stream of scientific literature must of necessity continue to grow so long as men continue to burrow into nature. Let us, however, dry up one perennial source of printed inanities by setting a right estimate on the minor question of date.

INDEX

academics 19, 99, 117
Acarus galvanicus 141
Adelaide Gallery 117, 127, 165
alkalines 51, 105, 111
alternate current 70, 189
ampère 206–207
Ampère, A.-M. 193
amusement 147, 149
anatomical science 59
anatomist 114
anatomy 56, 114
anxiety 7, 11, 81, 87
Arago, F 54–55
armature 189–191
Aron, H. 203, 205–206
Aronometer/Aron meter 203, 205–208
atom: chemical 75, 77–78; electrical 77; elements of 153–154, 156; isolating 154; potential force of 158–159
atomic: matter 151, 153, 156; theory 154–156; weight 156
attraction 21–22, 51, 53, 77–78, 195, 200, 205
Aurora Borealis 35, 54, 147
authority 59, 83, 85, 128, 154, 174, 196
axial 50–51, 53–54

Babbage, C. 3–5
Bain, A.: inventions of 117, 119; letter to Wheatstone 165–167
Barwise, J. 117, 119, 121, 127–128, 161, 163, 167
battery: cells 131, 144; Galvanic 136, 140n3; voltaic 141, 144, 190
Becquerel, E. 52
Bell, A.G. 72, 93, 211
Bentham. J. 47
biology 35, 63
Birkbeck, G. 117, 124, 166
bismuth 51–52
blood 25, 44, 51, 56, 60–61
Boyle, R. 21–22
Boys, T. 147, 149
British Association for the Advancement of Science 75, 127, 141, 159, 202
British empire 35, 211
Bryozoa 58
Bywater, J. 19, 21

Campbell, J. 129
caoutchouc (rubber) 51, 169, 176, 178
carbon rod 185
carbons 171, 175–176, 183–185
Carey, G. 93
carriage 117, 119
Chamber's Edinburgh Journal 35
chemist 75, 117, 155
chemistry 55, 64–65, 78, 147, 153–154, 157
Christian: altruism 81; era 21
chronometer maker 119, 127
clock: electric 72, 117, 125–126, 130, 161, 166–167; movement 122–123
clockmaker 117, 161, 165
coal: mines 13; substitutions for 11
cobalt 51
coil: of inductorium 191–192; magnet 190; of wire 69–70, 144
colonialism 84
Committee of General Literature and Education 35
commodity 5, 67, 203
communication: enhancement of 39; technology 93; telegraphic 121–122, 124
conduction 53, 153–154, 157–159, 200

consumers 5, 19, 169
controversy 117, 161, 167n1
Cooke, W. F. 117, 119, 135, 163–164
copper: piece 102; plated 99, 105, 137; weight 136; wire 69, 131, 136, 176, 189
cork 25, 122
Cowper, E. 128–129
Crosse, A. 141, 143, 145
Crosse, Cornelia A.H. 141
Cruikshank, G. 87, 90
Crystal Palace 67, 73
crystalline 52–53
crystallography 153
culture 55, 87
current: alternating 70, 75, 79, 205, 207–208; direct 169, 203, 206–208; electric 11, 52, 69, 71–72, 99, 107, 110–114, 123, 130–131, 134–136, 147, 167, 187, 198, 200–201; induction 52

Daniell, J.F. 117, 119–120
Darwin, Charles 29, 81
Darwin, Erasmus 29, 31
Davy, H. 6n2
defect 59, 65, 121, 207, 214–215
Deslandes. J. 182–183
dexterity 42, 89
diamagnetic 51–54, 196–197
disease 61–64
Ducommun foundry 169, 172
Dynamical Theory 195–196

economic divide 39
efficiency 14–15, 169
effluvia 21, 114
electric: ball 150; clocks 72, 117, 125–126, 130, 161, 166–167; fluid 22–23, 101, 104–114, 123–124, 130, 140n1; kite 149; shock 25–26
electrical: emission 22; engineering 93, 193, 203, 211–212; experiments 19, 24, 99, 147, 202–203; fountain 149; physics 187, 193; science 19, 35, 117, 122, 131, 135, 151, 187, 203; theory 151
electrician 19, 21, 25–28, 73, 78, 137, 141, 187, 193
Electrician, The 211, 213
electricity: active 37, 113; atmospheric 28n1, 51; conduction 53, 131, 153–154, 157, 158–159; discoveries in 21–23, 28; excitement 7, 35, 67; source of 135, 167; study of 29, 75, 77, 99
electrification 7, 149, 199, 211–212
electric power: discovery 22, 67, 71, 81, 104, 141, 203; future of 7, 39, 75, 93, 147, 187; systems 7
electromagnetic: attraction 195; current 52, 70, 191; field 193, 195–196, 198, 200–201; motor 147, 151; power 164, 189–190, 213
Electro Magnetic Clock 119, 121, 123–124, 137, 140n1
Electro Magnetic Printing Telegraph 119, 121, 123–124, 129, 140n1, 161, 164–165
electrostatic field 78, 195, 201, 203, 206, 217
energy 134; concerns 11, 151; dissipation 25, 28, 71; domestic 11; forms 77–78, 197; source 11; titans 7, 43
essay 19, 44, 54, 63–64, 87, 117
experiments: charging water 24–25 (*see also* Leyden); effects of lightning 27–28; electric circuit 26, 136, 147; errors 52; on gases 51, 53; of insects 143–144; materials used 24; methods of 19; researches 49–50, 151; with sealing wax 22–24
expertise 3, 161
Exposition Internationale d'Electricité 67

facsimile 95, 126
Faraday, M. 35, 47, 49–54, 69, 75, 117, 147, 151, 160, 187, 193, 197, 202, 214
Finlaison, J. 117, 121, 123, 128, 134–137, 140n1, 161
Fitzgerald, G. 93
Fontaine, H. 169, 171
force: centres of 157–159; electromotive 71, 191, 198–199, 201; *magne-crystallic* 53; of nature 41, 44, 51, 154; *optic-axis* 53; of water/steam 42, 67
forests 13–14, 215
Franklin, B. 19, 27–28
friction 15, 21, 114–115, 137–138, 173
fuel: alternatives 13–15; consumption 138, 173

Galvani, L. 29, 75, 77, 99, 111
Galvanic chain-work 29, 31
galvanometer 131, 133–135, 141, 190
Galvanism 99, 110–111
Garvey, M.A. 39, 41

gases 51, 53, 153
gender (female) 58
generator: electric 173, 187; friction 15, 21, 114–115, 137–138, 173, 199
geology 35, 64
George, H. 85
Gilbert, W. 21–22
glass globe 22, 24–25
gold 14, 113, 155
Gooday, G. 19, 147
Gramme, Z. 169
Gramme Dynamo/Gramme Machine 169, 172–174, *174*, 175, 178–179, 182–186
Grant, R. 55
gravity 51, 155–157, 205–207
Grey, S. 22–25
Grove's battery 131, 133

hair lines 23
hairspring 42
Harris, W.S. 117, 119
Hawksbee, F. 22
hemp/hempen line/sting 22–23, 27, 176
hempen line 22–23, 27
Henfrey, A. 55, 60
Henry, Joseph 93, 213–214
Henry, M. A. 213–214
Hertz, H. 78, 93, 193
Higgs, P. 169, 171
Hopkins, E. 55
Hughes, E. 55
human: productivity 39; purpose 14, 42, 45
Huxley, T. 47, 57–58, 84
Hyde Park 131, 135
Hydrozoa 57–58

imaginaries 93, 147
imagination 32, 39, 47, 54, 87, 122, 158
inefficiency 138
injunction 127
inking apparatus 125, 164–165
innovation 3, 4, 11, 211–212
intellect 6, 19, 21, 44, 47, 54, 63, 215
intelligence 42–44, 64, 78, 117, 119, 165
invention: challenge to 121–123; of electric telegraph 119, 124, 161; improvements to 127; ownership rights 125; rough model 126
inventor: first 140n1, 161; independent 123; rights of 128
Inventors' Advocate 161, 164

iron: blocks of 42, 52, 198; salts of 51; sulphate 53, 145; tube works 183
ivory 22–23, 51, 150

Jevons, W.S. 11, 13
Jones, R. 55–58

King's College 120, 124, 126, 135, 164–165
knowledge: human 43, 58, 62; mechanical 161; scientific 3, 4, 21, 141
Kölliker, A. 57

lamp: Serrin 172–173, 182–183, 185; suspension system 176, *177*, 178
lantern 176–177, 181–183
Laurie, P. 124, 166
lead 49, 104, 155
leather 51
Leuckart, R. 57
Leyden: experiment 25–26; flask 101–103, 115; Jar 78, 87, 99, 150, 199
light: bodies 22; ceiling 171, 174, 175, 179, 181, 184; electric 171–172, 174, 179, 185–186; expenses 182; lighted spaces 181–182
lightning: death from 28n1; effects of 27–28, 42, 54, 150
Literary Gazette 123, 129, 161, 163, 165
locomotion by steam 43
luminous: light/rays 179, 181, 184; vibrations 197, 202
Lunar Society 29

magnet: poles of 50–53; properties 187, 198, 200 (*see also* electro–magnet); strength 50
magnetic: action 52; poles 50–52, 198–199, 201; properties 51, 187, 200
magnetization 49–50, 197, 200
magnetism: augmentation 190–192; residual 190
manganese 51
Mann, R. J. 55
Marne and Rhine Canal 185
Martin, J. 164
mathematician 117
Maxwell, J.C. 93, 193, 195
McDowall, C. 123–124, 128
Mechanics' Magazine 55, 124, 161, 166
medicine 7, 25, 63–64, 114

Ménier Caoutchouc Factory 169
Ménier system of lamp suspension *177*
Ménier, Émile-Justin 169
Ménier, M. Henri 176, 178–179
merit 5, 63, 84, 121, 123, 127, 129, 136, 165
metal: common 51; negative 134, 136; positive 136–137
meteorology 29, 54
Mill, J. 47
Miller, W.A. 164
mirror 45, 49–50, 177
mismanagement 3
modernity 39
modification 69, 125, 165
morality 29, 39, 54, 63, 84, 87, 147, 161
Morse, S. 211
Moseley, H. 117, 119–120
muriatic acid 143–144
Muse 31–32
museum 87

National Debt Office and Government Calculator 121, 140n1
natural philosopher 19, 29, 99, 117, 141, 147, 151
networks 117, 141
Neumann, C. 195
Newton, I. 128
Newtonian theory of gravitation 215
nickel 51
nitrogen 51, 156
Northern Railway Company 179

Ohm, G. 71, 147
ohms 71
olive oil 51
Ørsted, H.C. 193
oxygen 31, 51, 156–157, 159

Paris Exhibition 67, 69, 72
Paris Observatory 55
patent: application 166–167; feuds/disputes 119, 130, 161; granted 163–164, 203
patentee 119–120, 213
pellicles 108, 114–115
Penny Magazine 35
perspective 3
petroleum 14, 179
phenomena: diamagnetic 196; electrical 7, 22, 47, 99, 111, 114, 136, 153, 195, 199–200; of light and heat 196–197; medusiform gemmation 57; social 83; thunder/lightning 54
Phillips, R. 141, 143, 145
Philosophical Radicals 47
Philosophical Transactions 115, 187, 195
phosphorus 14, 51, 114
physicist 7, 75, 213
physics 35, 49, 55, 64, 147, 157, 187, 193
physiologist 114
Pinkerton, R.C. 124, 128
Place, F. 47
plates: metal 99, 102, 104, 108–109, 113, 135–136; power 25
platinum 137, 155, 190–191
Plücker, J. 52–53
Poggendorff, J.C. 52
polarity 52, 189–190, 198–199
polarization, plane of 49–50, 197–198
politics 7, 87
Pollock, F. 129
Polytechnic Institution 117, 119, 121, 130, 165–166
portable engine 14, 109, 121, 184–185
Portsmouth 121–122, 140n2
potash 143–145, 156, 159
potassa 156
potassium 155–157, 159
practitioner 62, 89, 114
prejudices 21, 62–63, 154, 165
pretender 117, 121
Priestly, J. 19
printing: apparatus 125, 129–130, 165; telegraph 119, 121, 123–126, 128–129, 161, 164–167
printing-machine 164, 166–167
Priority 211, 213
probabilities 121, 155
productivity 39, 211
professional 72
Punch Magazine 7, 9, 87, 89, 91, 93

railway 44, 71–72, 86, 117, 119, 136–137, 179
reflection 3, 4, 49, 54, 114, 178
reforms 3, 35
Reich, F. 52
repulsion 22, 24, 51, 53, 195
resin 23, 29, 31, 51, 114–115, 155
resource(s) 3, 7, 11, 15, 39, 211
rheomotor 189–190
rhetoric 47

Robertson, J.C. 119
rotation: armature 189, 197; magnetism 54, 197
Royal Institution 19, 75, 151, 153
Royal Society 19, 29, 75, 99, 101, 123, 127, 151, 161, 163, 166, 187, 195

salts of iron 51
satire 7, 87
Saturday Magazine 35, 37
Sautter-Lemonnier and Co.'s Workshop 174–175, *175*
science: advancements 3, 15, 47; decline of 5; field of 27, 29, 50, 60, 75, 77, 79, 81, 84; progress 41–42, 55
sealing wax 22–24, 51, 155
Senlecq, C. 93
Siemans, C.W. 187
Siemans, Werner 187
signalizing telegraph 130
Silent Revolution, The 39, 41
silver 26, 101–106, 109, 111–113, 144, 155
Skating Company of Vienna 185
social change 39
Society for Promoting Christian Knowledge 35
Society of Civil Engineers 127
solids 51, 157–158
spinning mills 178–179
spiritualism 7
Stannius, F.H. 56
starch 51
static/statical electricity 79, 192, 195
steam: engine 6, 13–15, 138, 173, 176, 182–183; power 67, 71
Stockhardt, J.A. 55
Stokes, G.G. 75
Stuart Mill, J. 47
Sturgeon, W. 135, 141, 143
sugar refinery 90, 173, 176, 178, 185
sulphur 14, 51, 155–157, 199
Szczepanik, H.J. 93, 95

Taylor, R. 151, 153
technology 75, 87, 93, 147, 211
telegraphy 7, 71–72, 79, 93, 95, 117, 161
terrestrial 14, 42, 44, 54, 136
thermodynamics 79
Thompson, J.J. 75
Thompson, S.P. 39
Thomson, W. 39, 196–197, 200, 203, 206
timber 13–14
time-keeping 122; *see also* clock
tin 24, 26, 101–102, 104–105, 155–155, 184
tinkerers 7, 117, 187
Treatise on the Microscope in its Application to Vegetable Anatomy and Physiology 60
turpentine, oil of 51
Tyndall Correspondence Projects 141
Tyndall, J. 47, 52–53

United States Patent Office 213, 214
utilities 203
utopian thought 39

Valery, S. A. 187
Van Beneden, P.J. 57–58
vibration 78–79, 89, 137, 173, 197, 202
Victorian Britain 7, 39, 47, 211–212
video 93
Vogt, Oskar 57
Volta, A. 99, 101
voltaic: battery 141, 144, 190; circuit 130, 134, 136; electricity 131, 134, 136, 143; piles 101, 141
Von Siebold, C.T.E. 56

Ward, F.O. 164
watchmaker 119–120, 122–124
water: atoms 156, 158; body of 37, 107, 109, 112, 130, 134, 136, 149; as conductor of electricity 130–131, 133, 135
Watt, J. 14, 117, 140
watt hour meter (pendelzähler) 203, 208
Webb, F.C. 67, 69, 73
Weber, William E. 52, 195, 201
Westminster Review 47
Wheatstone, C.: letter to Bain 163; pretender 117, 121
Wheeler, S. 22–23, 25
Wilde, H. 187, 189
Willis, R. 117, 119–120
Wilson, S. 55
Winckler, H. 24–25
wood: as non-conducting 51, 105, 108, 134; insect experiment 144–145
Wright, T. 119, 121, 129–131, 135–136

zinc 99, 101–106, 109, 111–113, 145, 155, 181